Carolina Machado, J. Paulo Davim (Eds.)

Innovation Management

Also of Interest

Series: Advanced Composites
J. Paulo Davim (Ed.)
ISSN: 2192-8983
Published titles in this series:
Vol. 3: Metal Matrix Composites (2014) Ed. by Davim, J. Paulo
Vol. 2: Biomedical Composites (2013) Ed. by Davim, J. Paulo
Vol. 1: Nanocomposites (2013) Ed. by Davim, J. Paulo/Charitidis, Constantinos A.

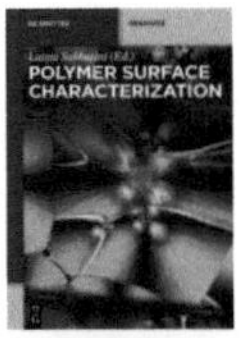

Polymer Surface Characterization
Luigia Sabbatini (Ed.), 2014
ISBN 978-3-11-027508-7, e-ISBN 978-3-11-028811-7

Nanomaterials in Joining
Constantinos A. Charitidis (Ed.), 2015
ISBN 978-3-11-033960-4, e-ISBN 978-3-11-033972-7

Functional Materials – For Energy, Sustainable Development and Biomedical Sciences
Mario Leclerc, Robert Gauvin (Eds.), 2014
ISBN 978-3-11-030781-8, e-ISBN 978-3-11-030782-5

Nanoparticles
Raz Jelinek, 2015
ISBN 978-3-11-033002-1, e-ISBN 978-3-11-033003-8

Nanocarbon-Inorganic Hybrids – Next Generation Composites for Sustainable Energy Applications
Dominik Eder, Robert Schlögl (Eds.), 2014
ISBN 978-3-11-026971-0, e-ISBN 978-3-11-026986-4

www.degruyter.com

Carolina Machado, J. Paulo Davim (Eds.)

Innovation Management

In Research and Industry

DE GRUYTER

ISBN 978-3-11-035872-8
e-ISBN (PDF) 978-3-11-035875-9
e-ISBN (EPUB) 978-3-11-038667-7

Library of Congress Cataloging-in-Publication Data
A CIP catalog record for this book has been applied for at the Library of Congress.

Bibliographic information published by the Deutsche Nationalbibliothek
The Deutsche Nationalbibliothek lists this publication in the Deutsche Nationalbibliografie; detailed bibliographic data are available on the Internet at http://dnb.dnb.de.

Cover image: Lite Productions/Thinkstock
Typesetting: le-tex publishing services GmbH, Leipzig
Printing and binding: CPI books GmbH, Leck
♾ Printed on acid-free paper
Printed in Germany

www.degruyter.com

Contents

Preface

Innovation Management in Research and Industry covers the issues of innovation management with a special emphasis in the field of research and industry. Markets, day after day, are changing more and quicker than ever. All over the world academics as well as practitioners are seeking to understand how organizations manage and/or can manage their knowledge and intellectual capital in order to establish and implement adequate strategies of innovation leading to more effective competitive advantages. This is a critical point not only in the academic and research field, but also in the industry arena where organizations need to develop and implement strategies able to facilitate and enhance innovation.

Conscious of the importance of these issues, and in order to answer the concerns expressed by many academics, as well as executives and managers, this book looks to help these professionals to understand and implement in their organizations effective strategies for innovation. Whilst focusing on these innovative strategies, this book is also concerned and interested in the application of theoretical concepts to modern organizations. It provides discussion and the exchange of information on strategies, techniques, methodologies and applications of innovation management in research and industry, as well as to communicate the latest developments and thinking concerning the latest research activity relating to these issues worldwide.

The book covers innovation management in research and industry in eight chapters. Chapter 1 discusses "*The Link Between Technological Focus and Innovative Capacities*". Chapter 2 covers *"Uptake of Market 'Induced Innovation' by Upstream Actors in Tanzania"*. Chapter 3 describes "*Customer Experience, Technology and Innovation: Evidence from Georgian London and the Victorian Era*". Chapter 4 contains information on "*Industrial Resilience: Reframing the Role of Innovation policies for Regional Development*". Subsequently, Chapter 5 covers "*Innovation Knowledge Management Nexus*". Chapter 6 discusses "*Human Capital Accounting: a Contribution to Innovation Management or a Fairly Tale?*". Chapter 7 describes "*Beyond the 3C Model in Collaboration Platforms: a Preliminary Analysis*". Finally, in Chapter 8, "*Emotion and Work: an Innovative Relationship?*" is presented.

For possible use in final undergraduate management and/or industrial engineering courses or as a subject example on management innovation at the postgraduate level, this present book also can serve as a useful reference for academics, researchers, managers, engineers and others professionals in related areas within innovation management in research and industry.

The Editors acknowledge their gratitude to De Gruyter for this opportunity and for its professional support. Finally, we would like to thank all chapter authors for their interest and availability to work on this project.

January, 2015

Carolina Machado, Braga, Portugal
J. Paulo Davim, Aveiro, Portugal

List of contributing authors

Neeta Baporikar
Scientific Research Department
Ministry of Higher Educatin
Salalah, Sultanate of Oman
neetajb@rediffmail.com
Chapter 5

Aykut Berber
Department of Management and Organization
Istanbul University School of Business
Istanbul, Turkey
berber@istanbul.edu.tr
Chapter 3

Filomena Antunes Brás
Department of Management
University of Minho
Braga, Portugal
filomena@eeg.uminho.pt
Chapter 6

Jairo Simião Dornelas
Federal University of Pernambuco
Administrative Science Department
Recife, Brazil
jairo@ufpe.br
Chapter 7

Ana Paula Ferreira
Department of Management
University of Minho
Braga, Portugal
aferreira@eeg.uminho.pt
Chapter 8

Dominic Hurni
Faculty of Business
Bern University of Applied Sciences
Bern, Switzerland
dominic.hurni@bfh.ch
Chapter 1

Christian Kunz
Chair of Business Administration and Controlling
University of Mannheim
Mannheim, Germany
christian.feuz@post.ch
Chapter 1

Pamela John Liana
The Open University of Tanzania
Morogoro Regional Centre
P. O. BOX 2062, Morogoro, Tanzania
pamela.liana@out.ac.tz
Chapter 2

Manuel Laranja
School of Economics and Management
Lisbon University
Lisbon, Portugal
mlaranja@iseg.ulisboa.pt
Chapter 4

Galinoma Lubawa
Institute of Rural Development Planning
P. O. BOX 138, Dodoma, Tanzania
glubawa@irdp.ac.tz
Chapter 2

Carolina Feliciana Machado
Department of Management
University of Minho
Braga, Portugal
carolina@eeg.uminho.pt
Chapter 8

Adrian Mäder
Faculty of Business
Bern University of Applied Sciences
Bern, Switzerland
maeder@liebefeld.ch
Chapter 1

Felix Adamu Nandonde
Department of Business and Management
Aalborg University
Aalborg, Denmark
fan@business.aau.dk
Chapter 2

Ana Lúcia Rodrigues
Department of Management
University of Minho
Braga, Portugal
alcrodrigues@eeg.uminho.pt
Chapter 8

Jorge da Silva Correia Neto
Department of Education
Universidade Federal Rural de Pernambuco
Recife, Brazil
jorgecorreianeto@gmail.com
Chapter 7

Andrea Gomes Santos
490, Couto Magalhães Street
Rosarinho
Recife, Brazil
andrea.g.santos@gmail.com
Chapter 7

Andreas Ninck
Faculty of Business
Bern University of Applied Sciences
Bern, Switzerland
andreas.ninck@bfh.ch
Chapter 1

Kim Oliver Tokarski
Faculty of Business
Bern University of Applied Sciences
Bern, Switzerland
kim.tokarski@bfh.ch
Chapter 1

Adrian Mäder, Christian Kunz, Andreas Ninck, Dominic Hurni, and Kim Oliver Tokarski*

1 The Link between Technological Focus and Innovative Capacities

Abstract: The common perception of innovation is colored by the mental image of high-tech laboratory facilities whose emphasis on cutting-edge technology is understood as proof of their ability to innovate. Does this perception reflect actual reality? This study uses expert interviews to operationalize the terms 'technological focus' and 'innovative capacities', whose presumed link is investigated in a series of qualitative interviews at companies in the service sector. The interviews reveal no immediate correlation between either variables.

1.1 Introduction

It seems a truism that the evolution of technology is important for innovation. The European Commission considers information and communication technology (ICT) in particular to be an important motor for growth and a source of innovation in the development of new products and processes [1]. Tschirky suggests that technological progress has gone beyond the realms of the manufacturing industry to leave its mark on the service sector, as evinced by the intensive reliance on ICT. Service sector companies are forced to engage much more intensively and on a strategic level with the fact that technology has become significant for their business [2].

A strategic confrontation with the issue of technology demands a more thorough investigation of aspects such as innovation and technology, without neglecting more traditional commercial concerns such as markets, personnel, or finances. In essence, a new means of measuring technology and innovation has to be found to understand whether there is indeed a link between the two. The following chapter considers the challenging proposition of operationalizing the two terms in the context of the service sector.

1.1.1 Purpose

This study explores the link between the intensity of the focus on technology and the ability of service sector organizations to innovate.

***Corresponding Author: Kim Oliver Tokarski:** Bern University of Applied Sciences/Haute école spécialisée bernoise, Faculty of Business, Institute for Corporate Development. Head of Institute for Corporate Development. Professor of Business Management and Entrepreneurship. Email: kim.tokarski@bfh.ch

The inquiry covers service providers covering a diverse range of businesses. In terms of technology, the study is limited to information and communication technology (ICT). Its objective covers three distinct research questions:

- Which criteria can be employed to establish the degree of technology focus at a company, and how can these be integrated into a technology focus rating scale?
- Which elements determine the innovativeness of a company, and how can these be translated into an innovative capacity rating scale?
- What correlation is there between a company's technology focus and its innovative capacities?

1.1.2 Study Design

Critiquing the assumption that a technological focus correlates directly with a company's ability to innovate requires the terms 'technology focus' and 'innovative capacities' to be meaningfully operationalized.

A survey of current literature was used to develop the tools explained in the following, that is, the 'technology focus gauge' and the 'innovative capacities maturity model'. Additional expert interviews were conducted to contribute to the 'technology focus gauge'.

The 'technology focus gauge' was used for the purpose of making a heterogeneous selection of research objects: service sector companies displaying different degrees of technology focus. Expert interviews were conducted at these organizations with the 'innovative capacities maturity model' providing the basis for the interview guidelines and evaluation matrix. The data was analyzed by a comparison of the independent variable 'technology focus' with the dependent variable 'innovative capacity' to elicit any potential correlation between the two.

1.2 Theory

Technology and innovation in the context of the service sector needs to be defined in detail to allow the correct operationalization of the terms. A possible means of distinction can be found in treating technology focus as an independent variable and its effect on innovative capacities as a dependent variable.

1.2.1 Services

A study of current literature on the concept of 'services' results in the following definition used in this study [2–5].

Services are activities,
- "that are based on potential capabilities which entail the ability and readiness to perform a service;
- in whose provision certain external factors are involved for whom or with whom the service is provided; and
- whose results represent certain material or immaterial effects on the external factors." [4]

1.2.2 Technology

Current literature also leads to the following definition of technology [6–9]:

Technology is the systematic, deliberate, and effective application of technological means to support or enable the provision of a service.

In terms of this definition, every company employs technology. The distinction lies in the strategic choice to make intensive use of modern technology or to develop such technology, which leads to the term 'technology focus'.

1.2.3 Technology Focus

Current literature tends to use the terms 'technologically focused' and 'technology-intensive' interchangeably [10–12]. However, the similarities are only skin deep, as a closer investigation reveals distinct differences:
- "A 'technology-intensive business' is characterized by the strategic readiness and ability to take in and develop new and novel production processes and products and thus achieve greater growth" [10].
- "Technology focus [refers to] the dominant place of technology in the business and all of its organizational units that wish to provide that technology. The dominant feature is therefore the presence of technological problem-solving skills, which are primarily found in research and development and in production" [13].
- Companies are considered technologically focused if "the purpose of their business consists primarily in selling goods and services that are based on utilizing new technological ideas, scientific advances, or systems, with considerable technical development typically being required before production can be started" [14].
- "Technologically focused companies are companies that are heavily invested in research and development activities (input-side innovation) and offer products that possess strongly novel features, that is, innovative content (output-side innovation). Innovativeness therefore refers essentially to the lasting commitment of companies to promoting or supporting the trying out of new ideas in the sense of having the abilities, opportunity, and readiness to do so" [15].

These definitions owe a debt to the industrial perspective and are not immediately applicable to service sector businesses. In this respect, Gelbmann and Vorbach's instrument for technology-oriented company analyses is used to define a meaningful definition of 'technology focus' [16]: In this sense, the "technological position of a company is determined by its internal technological resources (personnel and their know-how, available technical facilities), the access to external sources for technology, and its current position in terms of competitive technologies."

Tschirky confirms this definition in that technological achievements are the product of transdisciplinary expertise. A prominent feature of the development of new technologies is the increasing interconnection of different technologies. For the purposes of this study, this concerns in particular the use of information technology in the service sector. Tschirky also emphasizes that modern service providers are intensively involved with the technologies that are relevant for them, for example by developing the right know-how in dedicated departments [17].

Gelbmann and Vorbach agree with Porter that all activities of a company relate to the use of technology and that any technological change has an impact on virtually all value-creating activities in terms of its competitiveness. Their conclusion is that technology can be used both to create new points of distinction in the sense of innovative or improved products and to reduce costs by optimizing production processes [16].

With these considerations in mind and with a view to the ambition to measure the degree of technology focus, the following definition is employed:

The degree of technology focus reflects how important technology is for a company to produce commercial performance. It is irrelevant whether technology is used in core or support processes.

The definition provides a generally accepted basis for the proposed technology focus gauge.

1.2.4 Gauging Technology Focus

There seems to be no universally accepted means of measuring an organization's technology focus. Two reasons come to mind. First, there are different opinions about what technology and technology focus is or could be. Second, there is the difficult problem of finding suitable indicators and the relevant thresholds for measuring actual technology focus [18].

The Organisation for Economic Co-operation and Development (OECD) has established a clear scale for manufacturing businesses, ranging from low, low-medium, high-medium, and high sectors [19]. No such scale is available for service companies. A relevant measurement instrument is developed here, based on a review of current literature and expert interviews.

1.2.4.1 Literature

The OECD names only five knowledge intensive service sectors [20]: post and telecommunication (often not separable from each other), finance and insurance, service provision (without real estate), education, and healthcare.

Concerning the service sector, technology-intensive elements (without further specification) are seen in data processing and databases, research and development in the natural, engineering, agricultural, and medical sciences, architectural and engineering firms, and technical, physical, and chemical research [21].

The allocation to one of three technological phases pacemaker, key, and basis technologies represents another promising means of distinction. Pacemaker technologies are found in development, key technologies in concrete products and competitive processes, and basic technologies in competitively indifferent products and processes [9].

Another approach is proposed by Thudium, focusing on technology and affecting the management of technology-oriented companies in three ways [13].

Managing the development and transfer of technologies: The company acquires, develops, and sells technologies and technology-oriented products and/or services (e.g. quartz watches, high-power lasers, vaccines, rolling mills).

Managing the application of technology in business processes: The company uses technology in research and development, production, and service processes (e.g. computers in simulations, automated production, IT networks in corporate organizations).

Using technology to support business management: The company uses technology to manage the business and its processes (e.g. Management Information Systems [MIS], ICT for internal coordination, commissions, and controls).

1.2.4.2 Expert Interviews

Expert interviews were conducted with academics and business practitioners, covering possible qualitative and quantitative criteria. These were the results:

- A majority of the interviewees were in favor of using quantitative measures to allow for comparative responses, i.e. "more than" or "less than".
- Research and development (R&D) investments are not recommended as criteria in view of the sensitive data in the area and the heterogeneous nature of the service sector companies in question.
- Access to data is the general challenge when using quantitative criteria.
- The interviews confirm the impression of the literature review that there are currently no established systems.
- The customer's perspective can be applied to assess a company's technology focus.
- The scaling should employ a more general, qualitative level than closely circumscribed quantitative indicators.

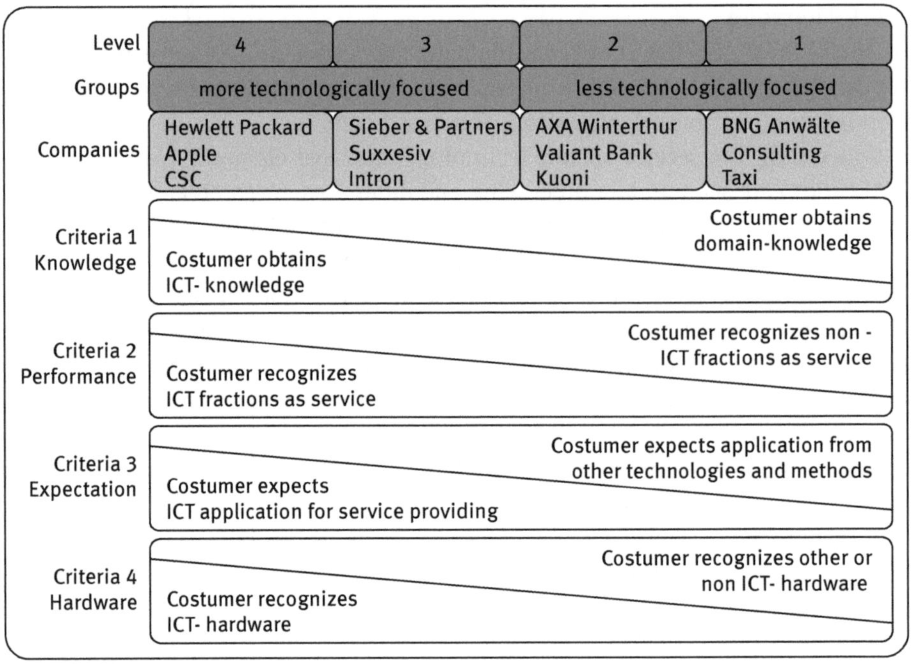

Fig. 1.1. Technology focus gauge

1.2.4.3 Gauging Tool

The technology focus gauge in Figure 1.1 was developed on the basis of the insights gained from literature and from the expert interviews.

The system proposed here for measuring technology focus distinguishes between four levels. Two levels each can be subsumed in the groups "more technologically focused" and "less technologically focused". This grouping shows that level 4 represents maximum technology focus and level 1 minimum technology focus. Specific example companies are allocated to each level to make the three following criteria more immediately comprehensible.

Criteria

Two levels were defined for all four criteria, with fluent boundaries, rather than a linear progression between them. The line is intended to show that there are no clear distinctions in this area. The criteria are not fully selective; rather, they add to and build on each other when it comes to allocating a company to one of the technology focus levels.

The first criterion "knowledge" includes the elements "ICT knowledge" and "domain knowledge". With increasing technology focus, the amount of ICT knowledge that the customers of a service company have also increases. In turn, an increase in the domain-specific knowledge occurs in companies with lower technology focus.

The second criterion "performance" covers "ICT performance" and "non-ICT performance". Consequently, the balance of the elements that form the actual performance for the customer changes across the four levels.

The third criterion "expectation" refers to the customers' expectations concerning the technologies and methods that the service company should employ when providing its services.

The fourth criterion "hardware" concerns the tangible elements of the services that the customer receives.

The following considerations and concrete examples help explain the gauging process in more detail:

Allocation

Customers of service providers on level 1 are not interested in the share of ICT in the provision of the service. For clients of solicitors, it matters that the litigation is successful and not whether the solicitor has gained his or her information from research in a library or in digital media.

Customers of level 2 companies, by contrast, expect a certain degree of ICT involvement. This can refer to e-banking services or web check-in for airlines. At the same time, the priority still lies on the actual financial transaction at the bank or the seats on planes.

Customers of service providers on level 3 demand services that consist mostly of ICT elements, for example because they do not have or do not want to acquire the necessary capabilities. This can concern IT system consulting or the coding of applications. One point of distinction from level 4 companies lies in the visible presence of ICT hardware. Advice on hardware aspects does not have priority in this type of service.

For customers of level 4 companies, the key distinction is the presence of hardware as part of the service. These services can, for example, be the operation of IT systems or the provision of broadcasting services via an internet infrastructure. For the customer, the hardware is an obvious element in the equation that cannot be taken away from the other parts of the service. In turn, it is less relevant for the client what the service provider could say about the TV shows it broadcasts or whether the service provider could develop the IT systems it hosts any further.

1.2.4.4 Assessing the Proposed Scale

The proposed scale essentially focuses on both the customer side of the equation and the core business of the company in question in the sense that both the providers and the users of services are often not interested in whether potential ICT elements are provided by the service provider itself or by a third party. The make-or-buy decisions of the service providers concerning these ICT elements do not therefore influence the perception of their technology focus. If this study had rated two or more insurance companies in terms of their innovative capabilities, they would have likely been awarded the same technology focus, irrespective of whether or not they host their own IT systems.

The chosen scale and its criteria also offer the opportunity to assess larger businesses in terms of their business areas.

Focusing on the customer's perspective has one essential advantage and one grave disadvantage. The advantage is that researchers and other interested parties can readily allocate the companies to the right place on the technology focus gauge. The relevant information is often readily available, for example in public domain documents about offered services. At the same time, the narrow focus on this one perspective can be unhelpful. Other qualitative elements, such as the ICT expertise of the companies' workforces, or quantitative elements, such as their involvement and investment into ICT, would add to the picture of the service providers' technology focus.

1.2.5 Innovation

When trying to see how the innovative capacities of a company can be assessed, a look at the terminology and possible classification criteria can offer some initial meaningful insights. There are as many attempts at defining and classifying the concept of innovation as there are books about innovation.

1.2.5.1 Novelty as the Basic Characteristic

Many publications consider Schumpeter the godfather of innovation, as in 1911 he described innovation (without using the term) in the sense of new combinations of goods, production methods, markets, or supply sources, but also new organizations [22].

Many attempts at defining the terminology rely strongly on the Latin origins of the term 'innovatio', which refer to "renovation, the creation of something new" [23]. For Bergmann and Daub, innovation is a catch-all term for anything new or improved [24]. Haber considers the quality of novelty as the key constituent trait of all definitions of innovation, although only innovations that are perceived as such by clients are of actual relevance [25]. Schmalen and Pechtl explain this novelty as the difference between the old and the new offerings in the markets. Novelty plays an important role as a factor determining an innovation's adoption on the demand side [26].

Hauschildt and Salomo name many attempts at defining innovation that relate to the aspect of novelty and consider this the terminological basis: “Innovations are qualitatively novel products or processes that are ‘noticeably’ different from the previous state” [27].

1.2.5.2 Classification

Innovation is seen from many different vantage points. A typical classification approach would distinguish between the contents, subjects, intensity, and success of innovations. [27–29]

Another dimension is the degree of involvement of external actors and thus the opening of the innovation process. The two poles of this dimension are open and closed innovation [30].

The innovation process represents the chronological dimension of innovation over time [22, 27, 29].

Conceptual Dimension

Hauschildt and Salomo distinguish between product and process innovations and state that innovation always represents a combination of both dimensions. In the case of service innovations, innovations to products and processes cannot be separated from each other, as this type of innovation does not imply a technological perspective. This relates to Schumpeter’s typology in that innovation can affect all functional aspects of business management and “the technical perspective is given up in favor of an administrative/managerial perspective” [27].

Subjective Dimension

The decision as to whether something is deemed innovative is, in the end, dependent on the number of people making that judgment. Interpersonal differences mean that the perspective that is applied by them becomes essential. Busse covers all possible subjects from the customers’/markets’ perspective and from the providers’ perspective [29].

Intensity

The dimension of intensity concerns the degree of novelty that the innovation brings. Current literature relies uncommonly heavily on the degree of innovation, either in dichotomous distinctions between radical and incremental, revolutionary versus evolutionary, or large versus small, or in multidimensional scales evaluating more complex degrees of innovativeness, as in Hauschildt and Salomon [27].

When using this dimension of intensity as a parameter for selecting the target companies for the study, the empirical link between their degree of innovativeness and their innovative capacities would need to be tested. If such a link is then established, the dimension ceases to be a suitable criterion, since the object of the inquiry, that is, innovative capacities, must not be influenced or postulated as an a priori premise.

Success

Another criterion can be seen in the success of the innovation. Since measuring success can be a complex endeavor with many problematic aspects concerning evaluation and judgment, this dimension does not seem suitable for the purposes of this study [28]. Another reason for its inclusion is named by Busse in that this dimension excludes future innovation in its limited retrospective focus on past successes [29].

Closed versus Open Innovation

Current literature on the open innovation theory tends to relate to Chesbrough's description of how organizational boundaries are made permeable for the (co-)development of innovation with external actors: "In the new model of open innovation, a company commercializes both its own ideas as well as innovations from other firms and seeks ways to bring its in-house ideas to market by deploying pathways outside its current businesses" [31].

For Reichwald and Piller, this opening becomes the starting point for interactive value creation, with interactivity meaning "the cooperation between companies and external experts, clients, and consumers in the value-adding activities during the innovation process [. . .]" [32].

As in the degree of innovativeness itself, the degree of the involvement of external actors needs to consider any empirical proof of effects on the innovative capacities. If such effects are indeed established, the dimension would seem unsuitable as a criterion for the purposes of the study, since they represent an unintended influence on the actual object of research, i.e. the innovative capacities (e.g. exploring only companies that have committed fully to open innovation processes).

Processes

A procedural perspective is relevant for the chronological definition of the term innovation. Depending on the author in question, the innovation process can cover several phases. This procedural dimension typically begins with the generation of ideas and ends with their launch in the market. This apparently generally accepted last step seems more important than the number of steps leading up to it: "There is no doubt that the innovation process has to cover at least the phases until the new product has been introduced in the markets" [27]. "A very narrow definition of the term innovation refers only to the process of establishing innovations in the market and thus to the point of their market launch [...]" [29].

1.2.5.3 Service Innovation

The terms 'service' and 'innovation' have been treated separately up to this point. An integrated assessment and definition is now required for the purposes of this study. The relevant dimensions 'potential', 'process', and 'output orientation' have already been discussed above; innovation relates to all three of these dimensions. This means:
- service potential innovation in the sense of the ability and readiness to produce a service;
- service process innovation linking existing and established processes or adding new processes;
- service output innovation focusing on the impact of a service's delivery on external objects.

1.2.5.4 Defining Service Innovation

The following perspective is applied in this study, distilled from the multitude of applicable definition and classification dimensions.
- The aspect of novelty is considered holistically. This aspect concerns novel products, services, processes, organizational structures, cultures, and business models.
- Since the degree of technology focus represents an explicit criterion in this study, the inquiry cannot limit itself to technological innovation alone.
- As the study is based on the experience and knowledge of experts in companies, it applies an internal, intraorganizational viewpoint from the perspective of the service providers.
- The procedural dimension reminds us that the perspective has to cover all steps from the generation of a novel idea to the launch of a marketable service. Innovation in this sense only refers to a novel service that has been actually established in the markets.
- The focus on services and on service innovation again points to the importance of an intraorganizational vantage point concerning the ability and readiness to produce a service, but also to the importance of the procedural dimension. In this respect, the involvement of external factors (actors) is essential.

Schniering has proven the significant and positive impact of innovative capacities on the commercial success of service innovations [28]. No further exploration of the prospects of success of an innovative service is therefore required, and the study can focus exclusively on the innovative capacities of the organizations in question.

1.2.6 Innovative Capacities

The ability to innovate represents the key object of interest in this study and, as such, the variable measured in the empirical inquiry. Innovative capacities are defined and considered from a variety of viewpoints. A review of current literature reveals a specific set of elements that are preferred models when explaining innovative capacities:
- resource-based view
- competence-based view
- organizational structures (ambidextrous organization)
- knowledge-based view (absorptive capacity)
- learning-based view (learning organization)
- path dependency
- process-based view (innovation processes)

These explanatory concepts are used as the basis for the design of a maturity model to measure the innovative capacities of organizations.

1.2.6.1 Definitions

There are as many definitions of innovative capacities as there are different approaches to the subject. Studies that address the decisive elements for innovation on a meta level tend to define the ability to innovate as a multidimensional and dynamic meta-competence covering a set of unique elements [33].

1.2.6.2 Resource-Based Views

Burr and Stephan postulate that innovative capacities are available when companies manage to establish new capabilities for producing novel services [34].

Many empirical studies investigating the success factors in innovation follow this resource-based approach. For instance, Schniering considers innovative capacities in terms of resource-based indicators, which in turn relate to the successful innovation of services [28].

Kühnl posits human resources as relevant indicators in his concept of internal adoption, again relating directly to successful innovation [35].

For Burr and Stephan, the resources in Table 1.1 are required for service innovation.

1.2.6.3 Competence-Based Views

The competence-based approach represents an extension of the resource-based approach. Hamel and Prahalad consider the concept of ‘resource leverage’ to be an important element of such a competence-focused view. “By sufficiently concentrating,

Table 1.1. Resources required for service innovation. Source: Author's own, following Burr and Stephan [34].

Category	Resources
Human assets	Professionally qualified and motivated personnel The ability to develop innovative services The ability to produce offers and implement them on site at the client's business
Reputation	The company is seen as the successful provider of innovative services A reputation that promotes the acceptance of new services
Brand	Protection from imitation and communication with clients
Organization	Efficient; development and innovation-oriented structures Planning, reporting, control, and coordination systems to support the development of innovations (e.g. project management tools, balance score card, service level management)
Technology	IT systems that support the prototyping of new services Communication and information systems (intranet, knowledge databases, newsgroups) IT systems for delivering the services in the market: – Online offerings (e-commerce, e-banking, e-insurance) – Communication platforms with the client (esp. for integrating lead users) – Mobile technologies (e.g. on-site sales support)
Culture	Innovation and customer-oriented behavior is rewarded Unplanned, spontaneous, and creativity-oriented practices are encouraged Problem and results-oriented actions and decisions are encouraged An open mind concerning ideas from outside the company is promoted Personal and informal communication and sharing of information is encouraged The failure of innovation ventures is tolerated (failure tolerance) Contradicting superiors is allowed or even encouraged
Finances *	Cash and available credit
Material assets *	Machines, facilities, real estate

* Factors of secondary importance

efficiently accumulating, creatively complementing, carefully conserving, and speedily recovering resources, firms close the gap between where they are and where they want to be" [36].

Competence-based approaches are subject to considerable debate and critical scrutiny. The key lies in the definition of competence. If competence is understood to mean self-organized problem-solving skills that are used to stabilize and maintain experiences and abilities, it is indeed of relevance for the challenge of innovation [24]. By challenge, we mean the traditional conflict between mechanistic and organic management systems. While mechanistic systems aim for the effective and efficient handling

Table 1.2. Concept for innovation-oriented organizations. Source: Author's own, following Hauschildt and Salomo [27].

Openness of the system	Permanent public accessibility of the organization Conscious interaction with opinion leaders Readiness for taking in and releasing information Readiness for a dialogue about innovation Receptiveness to impulses and change
Degree of organization	Definition of a degree of organization with as few rules and prohibitions as possible Organization meaning freedom to act without constraints
Style of information	As little regulation of the flow of information as possible Promotion of informal communication channels Discussion and negotiation of innovation as important as ad-hoc issues or routine problems
Promotion of cooperation	Readiness for cooperation and encouragement of all involved and affected parties Mutual appreciation of different departments and functional areas
Perception of conflicts	Conscious acceptance of conflicts as a source of creativity Constructive attitudes about conflict
Recruitment and HR development	Demand for unconventional personnel who are able to produce and manage conflict Competences for solving problems Ability to assert new ideas
Competence and responsibility	Flexibility in the allocation of responsibilities Freedom for innovative pursuits outside of routine duties Explicit encouragement of innovation initiatives

of repetitive activities, organic systems respond to a changing environment with ambiguous and changeable targets. With this in mind, Hauschildt and Salomo call for a combination of both systems in an innovation-oriented organizational concept with the focus elements in Table 1.2.

The ability to adapt to a changing environment is the primary point of criticism in traditional competence-oriented approaches [37]. A dynamic alternative is discussed in Section 1.2.6.4.

When trying to allocate competences to different levels, Freiling et al. mention ways to refine goods, promote competences for engaging with the market, and guarantee new development as part of their meta-competences (Table 1.3).

Sammerl distinguishes between two degree of competence. The first degree concerns everyday performance, whereas the second addresses the meta-ability to coordinate and develop resources effectively [33] (Table 1.4).

Table 1.3. Different levels of competence. Source: Author's own, following Freiling et al. [38].

Refinement competences	Refining of input goods in line with the resources currently in demand Developing currently available input goods, resources, and competences (also referring to the available decision-making authority and creative freedom) Having access to external input goods, resources, and competences
Market interaction competences	Activating the readiness to produce/provide a concrete service at a specific time Conducting successful market transactions
Meta-competences	Anticipating future requirements of the system's environment and proactively shaping future performance potential Ensuring permanent adaptation and change abilities (flexibility) Ensuring deutero learning Being aware of change underway in the environment

Table 1.4. Two degrees of competence. Source: Author's own.

Degree 1	Alignment with the stabilizing system (maintaining experience and expertise) Functional competences and core competences for producing routine performance with the available means (refining competence) Low maturity
Degree 2	Potential of the company to evolve and adjust to its environment Dynamic outlook (dynamic capabilities), meta-competences High maturity

1.2.6.4 Ambidextrous Organization

The St. Gallen Center for Organizational Excellence (CORE) conducted a longitudinal study with the 300 largest European enterprises between 1995 and 2004. One basic result was the insight that companies enjoying long-term profitable growth have focused their organizations on efficiency and flexibility [39]. This dual focus is described as organizational ambidextrousness. The study revealed four ambidextrous strategies: (1) cyclical change, (2) physical separation, (3) parallel organization, and (4) integrated networks.

For this to succeed, a holistic approach is decisive as it aligns such often disparate elements as the company's culture, its leadership structure, and its HR system (Table 1.5).

Culture and (lateral) leadership are closely intertwined with the idea of (ambidextrous) organizations. They are important levers in managing the inherent contradictions. A closer analysis should check their potential application in the assessment of an organization's innovative capacities.

Table 1.5. A holistic approach for an innovation-oriented organizational structure. Source: Author's own, following Gomez et al. [39].

Organizational strategy	Conscious choice and use of a suitable ambidextrous strategy (cyclical change, physical separation, parallel organization, integrated network)
Corporate culture	Formation of unique cultural traits in different business units to respond more flexibly to individual customers' needs Prevention of silo thinking and promotion of a sense of identification with the company as a whole (e.g. with 'one brand' programmes) Measures to promote a shared identity (one brand, interdepartmental networks, IT platforms, training, workshops)
Leadership structure	Visionary leaders with creative ideas for new products and business opportunities Pragmatic leaders with implementation skills Promotion of diversity Diversification of the traditional focus with team work and cooperation models
HR systems	Rewards for departmental performance and interdepartmental cooperation Promotion and support for managers engaging in lateral cooperation

Culture in Ambidextrous Organizations

The term culture is often the subject of controversial debates where innovation or the need for change are concerned. Schreyögg mentions the negative effect of corporate cultures on organizations' ability to adjust. Clear guidelines, homogeneous frames of reference, consistent decision-making processes, and the motivation and esprit du corps promoted by clear values are essentially positive aspects of a strong culture. At the same time, they can promote rigidity and undermine flexibility. "Barriers to change are created when the members of an organization act in accordance with their accustomed frames of reference, values, and rituals" [40].

This seems to contrast with the everyday experience of innovation-oriented cultures. Schreyögg emphatically denies such an orientation, since innovative elements are inherently opposed to strong cultures. Strong corporate cultures that are constantly in flux are a contradiction in terms: "Systems in constant change cannot have a strong culture in the true sense of the term, since they do not follow any set norms" [40].

This suggests that strong corporate cultures cannot have a positive effect on the ability to innovate. To create enough leeway and freedom for new thinking, Schreyögg suggests escaping the shackles of a strong culture [40].

Lateral Leadership in Ambidextrous Organizations

By lateral leadership, Kühl and Schnelle mean the ability to establish cooperation processes without formal authority over other people. The need for such lateral leader-

Table 1.6. Elements of lateral leadership. Source: Author's own, following Kühl and Schnelle [41].

Mediation	Status quo	Rigid patterns of thinking and explanations are recognized
	Trick 1	The situation is recast by projecting it into the future Speculative questions: "What would happen if?"
	Trick 2	Incompatible axioms are applied by engaging with contradictions
	Trick 3	Paradoxical interventions are made by artificially exaggerating current patterns
	Trick 4	Irritation is created by producing instances of discrepancy
	Trick 5	Partial mediation is accepted in favor of a full consensus
Power	Status quo	The current distribution of power is analyzed Power is understood as asymmetric interactions
	Option 1	A space is created for the negotiation of power The lateral leader promotes the negotiation of power A mediation process is introduced
	Option 2	The distribution of power is changed by introducing new actors Backup is brought in
	Option 3	Available options are changed and shared by creating relevant exchanges
Trust	Risk	Trust can be abused (social transactions) Trust quickly turns into distrust
	Analysis	Possible 'battlefields' for creating trust are identified
	Option 1	The stakes are lowered
	Option 2	Objective constraints are disclosed
Balance	Mediation, power, and trust processes are always happening in parallel Mediation, power, and trust processes can support, inhibit, or replace one another The optimum balance depends on the given situation – there is no single *correct* balance	

ship arises from so-called local rationalities, created by the diversification of functions or product units. The intended effect is the development of the ability to master highly complex challenges [41]. In ambidextrous organizations, these local rationalities are formed by flexible structures that promote innovation. For Kühl and Schnelle, mediation, power, and trust represent the means of control in lateral leadership. In the end, the right balance of these elements is key. Table 1.6 brings together the various mechanisms by which these elements operate and helps establish whether lateral leadership is actually in place.

Absorptive Capacity

Cohen and Levinthal proposed the absorptive capacity of organizations as a key factor for successful innovations in 1990: "The ability of a firm to recognize the value of new, external information, assimilate it, and apply it to commercial ends is critical to its innovative capabilities" [42].

Schreyögg relates the recognition, assimilation, and application of sources of knowledge to the organization's learning track record. The development of knowledge only becomes available for organizations if they work permanently on their absorptive capacity. The poorer that ability, the shorter the reach of their innovations, which means that improvements are eventually only looked for in the organization's closely circumscribed and familiar area of activity [43]. The company's behavior becomes path-dependent, i.e. it unintentionally restricts itself more and more to patterns of perception and practice that have formed in its history. [43]

The operationalization of this factor for the purposes of this study needs to expand the restrictive focus on R&D to encompass the organization as a whole. In this respect, the absorptive capacity of an organization is a procedural ability that is expressed in established routines in the sense of automatic patterns of behavior and absorptive practices. Schreyögg names specific sub-competences for each phase (identification, assimilation, and implementation) that can be used to ascertain the actual absorptive capacities of a company (Table 1.7).

Table 1.7. Specific competences in the absorption process. Source: Author's own, following Schreyögg [43].

Identification (Acquisition)	Assignment of formal and informal gatekeepers as 'integrating elements' Presence of functions with cross-border relevance Support for partners' aptitude for learning by cooperating with external partners (e.g. involving lead users in the development process) Ensuring the diversity of perspectives to recognize acquisition potential with a broad foundation in the organization (e.g. recruitment, development, and promotion) Avoidance of a too rigid corporate culture that over-intensifies pride in past achievements
Integration	Will to distribute new knowledge Avoidance of silo thinking Regular contacts with other business areas Work in cross-functional project teams
Implementation (Exploitation)	Use of change agents to support the changes Self-critical analysis of past learning and exploitation processes Application of reflective practices (feedback sessions, confrontations, quality circles) to promote the (deuteron) learning process

Learning Organizations

The ability to learn means the ability to innovate: "The innovative capacities of organizations depend [...] strongly on individual's ability to learn" [24].

The relationship between learning and not-learning is particularly relevant where ambidextrous organizations are concerned. "Organizational learning [...] also implies

[…] intended not-learning, that is, the decision to not learn and not change in certain circumstances (stabilization). A learning system also needs to learn how to use the advantages of not-learning" [44].

The idea of the 'learning organization' has been hotly debated and interpreted in many different ways. Older concepts in didactics consider individuals and their experience-driven, adaptively rational learning processes as the engine of organizational learning [45]. Argris and Schön treat the knowledge of organizations in the form of action and practice theories that are unique to each organization; the organization here acts as a single entity [46]. Such organizational practice happens on three levels (Table 1.8).

Table 1.8. Levels of Learning. Source: Author's own, following Argyris and Schön; Steinmann and Schreyögg [45, 46].

Single-Loop Learning	Regulating cycle that corrects deviations from the target state Instrumental learning for better operational performance The basic values, norms, and strategies of the company remain unchanged
Double-Loop Learning	The basic values, norms, and strategies of the company are challenged Learning leads to a change of values and guiding theories and strategies The target state itself is challenged Need for open and unbiased mindsets in the organization
Deutero Learning	The third level is a meta-level of organizational learning Reviewing past learning processes is meant to produce continuous readiness for learning

Organizations will become most effective at learning by reflecting on their own learning processes. Single-loop learning cannot constitute a learning organization in this respect. However, more recent concepts have been criticizing this point. From the point of view of systems theory, organizations have been learning ever since they came into being: "Learning and developing or not learning and being unresponsive to certain impulses for change are two types of abilities that are equally important for the survival of organizations" [47]. This (natural) evolutionary learning mechanism forces us to consider whether organizations can access their own evolutionary mechanisms, i.e. learn how to change their way of learning. For Wimmer, the actual learning process is dependent on irritation at the place where systems and their environments meet. Wimmer defines eight success factors that have a positive effect on an organization's susceptibility to irritation (learning impulses) and ability to process them (Table 1.9).

Table 1.9. Factors for managing irritation successfully. Source: Author's own, following Wimmer [47].

Degree of environmental sensitivity (External contacts)	Intensity of cooperation with the client (shared acquisition of know-how, processing of feedback, learning from customers' problems) Monitoring of the markets and development beyond existing customers (industry analyses, economic data, market research institutes) Cooperation with suppliers in the value chain (supplier integration, business networks, loose cooperation) Constant monitoring of competitive dynamics Benchmark as incentives for development (instead of simple models to be copied)
Knowledge management (Knowledge base)	Generation of implicit knowledge by acting personnel Dependence on holders of knowledge = intelligence of the organization (application and sharing) Management of the generation of information in the organization (accidental versus planned) Internal openness for sharing knowledge (culture of learning from each other) Openness for external influences Handling of the tension between maintaining (old) knowledge versus generation new solutions for problems (rules)
Manner of failure management (Deviation management)	Type of error culture (avoiding a 0% error culture or perfectionism) Culture of continuous improvement, instead of error avoidance Promotion of a climate for experimenting and taking risks Freedom to fail and effective learning from experiences Intelligent questions, instead of fast answers
Maturity of leadership structures (Cooperativeness in the leadership team)	Leadership constellation receptive to impulses for learning Clear distinctions between organizational levels with mutual respect for different responsibilities Cooperation and decision-making ability of the leadership team and cooperation across levels Conflicts as opportunities for creative problem-solving processes
Type and extent of cooperation (Extent and quality of cooperation)	Integration of highly specialist areas of know-how (occasions for interprofessional cooperation) Integration of specific disciplines but mandatory search for interdisciplinary solutions Simultaneous cooperation of all areas in new developments (experts, production, sales, customers, development) instead of sequential problem-solving Promotion of forms of working that promote learning (teams, group work)
Self-reflection (Time as a crucial resource)	Balance between operational acceleration and slack time for shared reflection Regular and high-quality target reviews Ensuring feedback for executives Auditing of management potential Reflective strategy workshops for regular status quo assessments
Innovation-positive HR management (Focus on the link between individual developmental potential and organizational innovative dynamics)	Focused recruitment and integration of holders of specific know-how Focused potential development Horizontal mobility (diversity of career prospects) Promotion of self-reflective learning formats Access to the implicit experience of holders of specific know-how Suitable calculation of compensation that considers the mobilization of knowledge
Problem attribution and management (Handling of problems in difficult situations)	Avoidance of externalizing problems or personal blame (Personalization makes organizations immune to change) Handling and treating difficult situations constructively (maintaining integrity in situations that can be embarrassing or threatening) Problems seen as opportunities for learning, instead of attributing lack of knowledge to personal incompetence Readiness to involve external persons (coaches) in difficult situations

1.2.6.5 Clients' Integration in Services

In the previous argumentation we mentioned the importance of using external interfaces. In addition to being intangible, a key characteristic of services is the involvement of the client, since services cannot be provided without the presence of an external entity (object or person). The client represents a valuable resource for the innovation process, as the company can tap into the client's knowledge [48].

Achieving this form of integration needs certain management and client integration skills. Vesshoff and Freiling define such client integration skills on several systemic levels, with the key implications in Table 1.10.

Table 1.10. Systemic levels of customer integration. Source: Author's own, following Vesshoff and Freiling [49].

Strategic Logics	Conscious management of different perceptions, judgments, and practices of providers and clients Establishment of conflict avoidance and management patterns Definition of a framework for conflict management Creation of a relationship-positive atmosphere
Management Processes	Combination of strategic logics and management processes: – Relevant coordination needs observed by management and included in the subsequent design of the value creation process – Impact on the provider's basic market approach – Focused investment as a result of innovation partnerships
Potential	Identification of suitable clients – especially lead users – and the relevant external factors Management of the diversity of external factors during temporary integration Coordination with internal potential Definition of the scope of integration (poles: 'design for the customer' versus 'design by the customer')
Processes	Provider's ability to absorb knowledge Development of partner-oriented routines to improve efficiency and effectiveness in the innovation process without falling into path dependency Use of professional and power multipliers to create motivational forces Transparency about processes and roles with early and open communication
Effects	Application of the results of the innovation process Adjustment and realignment of service provision processes as a condition for the later integration of clients

1.2.6.6 Path Dependency

Path dependency refers to practices and expectations stuck in the proverbial rut that force managers to act against their better instincts [50]. This effect occurs when the people involved lose the room and freedom to step back and think in their everyday work. Focusing on efficiency means less consideration for effectiveness (Are we producing what the markets want?). Breaking free from these paths is only possible if such instances of recognition are actually translated into real action, for example in the sense of individual targets implemented to achieve a new strategic direction.

Dievernich points to performance reviews as an instrument for organizational practice that should cover both the short-term, efficiency-oriented component and the long-term, innovation-oriented dimension: "Executives need to comply with the need for more short-term efficiency (single-loop learning) and pave the way for options that can only come true in the later future (double-loop learning)" [50].

1.2.6.7 Innovation Process and Implementation Competence

The ability to innovate as a meta-level competence as defined in this study also includes the ability for its practical implementation as a decisive element in the process: "The success of any innovation stands or falls with the structure and realization of the development process" [52]. In addition to the ability concerning the conscious swinging and switching between the phases, professional systems and methodology are key factors for the successful completion of all phases of the innovation process. Acquiring such methodological and systemic competences is an "indispensable precondition for the high-quality implementation of the innovation process" [22].

Stern and Jaberg emphasize the organization of the project (or its characteristic success factors) for the effective realization of innovations [52]:
- interdisciplinary teams
- project organization and project management
- available resources and responsibilities in the team
- staffing and cooperation; motivation, types, roles
- application of development techniques

As the interdisciplinary aspect (ambidextrous organizations) and the question of resources (resource-oriented approach) have already been analyzed in detail, the focus now lies on the aspects of project organization, project management, and team structure. The most important implications are summarized in Table 1.11.

1.2.7 Innovative Capacities Maturity Model

The analyzed aspects of innovative capacities are now classified into distinct degrees of maturity. Each level represents a defined level of innovativeness in the company in

Table 1.11. Factors for successful process organizations. Source: Author's own, following Stern and Jaberg [52].

Project organization and milestones	Regular checking of progress as defined points (milestones) Objectiveness and self-criticism as a condition for decision-making skills (e.g. stopping projects early if need be) Shared definition of targets with functional specifications (turnover, costs, timing, market, and customers) Requirements and specifications for technical implementation (use cases, production planning, technical specifications, timing and budgets, marketing and sales plans) Tests (usability tests), prototype construction for acceptance tests
Project management: Techniques and controlling	Constant tracking of time, costs, and conceptual targets Definition of the optimum timing for market entries Constant identification and monitoring (network analysis and Gantt charts) of the parameters in the development process Awareness of the importance of time-to-market Application of suitable techniques for maximum the pace of development (parallelization, flexibility, teamwork, communication, simultaneous engineering) Definition of the right pace (daring to slow down) Risk management, risk portfolio, project monitoring (project FMEA)
Team staffing	Optimum team structures and cooperation on the basis of a team culture Awareness of the great importance of soft skills (motivation) Definition of specific key roles in the innovation process: – Functional multipliers (functional development competence; passion and commitment) – Power multipliers (support for innovation from top management) – Process multipliers (methodical, mediation, and integration skills)

question. A higher degree can only be achieved if the requirements for the previous degree are fulfilled completely. This evolutionary model rates the innovative capacities of the participating companies from 1 (the lowest degree) to 5 (the highest degree). Every degree is a multidimensional construct, conceptualized with the relevant variable, indicators, and key questions described in the following.

1.2.7.1 Conceptual Definition of the Degrees of Maturity

The variables themselves represent complex forms that cannot be measured with single dimensions. Suitable indicators are chosen for each variable for the following application in the empirical study. These indicators reveal the aspects for which information needs to be sourced [53]. At the same time, the definition of the indicators also implies the definition of possible measuring criteria for assessing the interviewees' statements in terms of their companies' fulfillment of the requirements.

1.2.7.2 First Degree of Maturity

The first degree of maturity revolves around the necessary resources that underlie innovation activities (Table 1.12). The focus lies on human resources, as people and their competences represent a specific ability to develop novel and innovative services. The key criterion is the ability to recognize and seize such potential capability. Another resource of relevance is the brand of the company in question. It is both an important tool in communication with the client and a public signal of the innovative capacities of the company.

In accordance with the understanding of competence employed in this study, this first degree of maturity is limited to competence degree 1: Competences defined in the traditional sense. Competence here focuses on existing elements of competitiveness, concerning functional competences and core competences that combine with the available resources to make everyday performance possible.

The first degree of maturity is expressed conceptually with the variables 'human assets' and 'brand'. The ability and readiness of employees, the recruitment of human resources, the value of the brand, and its link with innovation represent the relevant indicators in this respect.

Table 1.12. First degree of maturity – variables, indicators, and key questions. Source: Illustration by the author.

Variable	Indicator	Key Question	Indicator Achieved
Human Assets	Capability and readiness	How do you rate the current situation in terms of the capability and readiness of employees to develop novel services?	Capability and readiness are named explicitly in the statement.
	Recruitment	What do you do to retain or recruit such personnel?	Innovation plays a direct or indirect role in recruitment; active measures are taken to retain innovative minds.
Brand	Value	How important is the brand in your company?	The brand's value is clearly expressed or the brand is used in communication.
	Link with innovation	How does the brand's perception relate to innovation?	The link is recognized, but not necessary used, since the general thrust might not combine well with innovation

Table 1.13. Second degree of maturity – variables, indicators, and key questions. Source: Illustration by the author.

Variable	Indicator	Key Question	Indicator	
			Achieved	Not Achieved
Organization	Dynamic forms of organization	Where, how, and when does innovation happen at the company? (Possible explanations: own department, innovation board, letterbox, process, etc.)	Specific locations, regular / planned activities, ambidextrous	Unknown location of innovation; unstructured handling of ideas
Leadership	Balance	How do executives deal with the tension between stability and change in the innovation process?	Bias towards flexibility; balancing, lateral leadership; focused efforts to maintain a balance recognizable.	Bias towards stability; coping with imbalance
Culture	Alignment with the corporate culture	How well does the culture of the company relate to innovation activities?	Traits like agility, flexibility, efficiency, and market awareness are visible despite the presence of traditional values.	Innovation is held back by cultural factors.

1.2.7.3 Second Degree of Maturity

The formation of organizational structures that promote innovation leads to the second degree of maturity (Table 1.13). In addition to the creation of suitable structures, this concerns in particular the response to the inherent friction of stability versus change. The relevant ability here concerns the ability to cope with this dynamic element, the transition from one form to another.

The development of flexible and ambidextrous structures produces locally specific rationalities. Frequently, executives do not have the necessary authority, but are forced to achieve a certain effect. Such pliant structures demand a lot from leadership, as it is a difficult balancing act to achieve. Executives with such lateral leadership skills therefore contribute massively to the ability of companies to innovate.

Corporate cultures are intrinsically linked with corporate structures. Strong corporate cultures can, however, introduce an element of rigidity and lack adaptability. Even the common parlance term of a 'culture of innovation' does not manage to reconcile the different elements of innovation and culture. For a company to be able to innovate, it needs to emancipate itself from any too-tight cultural corset.

The second degree of maturity is conceptually expressed in the variables 'organization', 'leadership', and 'culture'. Dynamic forms of organization, the search for the right balance of stability and change, and the degree of reliance on the corporate culture are the relevant indicators.

1.2.7.4 Third Degree of Maturity

The essential element of the third degree of maturity is the company's ability to learn (Table 1.14). The structural elements covered by the third degree allow the organization to develop certain absorption processes that can prevent a too rigid corporate culture or instances of silo thinking. These absorption processes give the company access to knowledge about events or needs in or from its environment. One important element of this receptiveness to outside influences is the integration of the client: Scouting, integrating, and implementing innovations in tandem with the client is a key motor in the innovation of services.

The degree of (external) receptiveness determines the company's sensitivity to and irritability by outside forces. It leads to a way of handling knowledge that allows a conscious distinction between learning and not learning. The company then has the ability to decide in favor of change or in favor of stability.

The company's ability to engage in self-reflection is closely linked to this aspect. This is only present if there is sufficient time for reflection and self-criticism. The greatest aptitude for learning can only be present if the company is able to reflect consciously on its own learning processes.

Table 1.14. Third degree of maturity – variables, indicators, and key questions. Source: Illustration by the author.

Variable	Indicator	Key Question	Indicator	
			Achieved	**Not Achieved**
Receptiveness	Sourcing of information	How does the company acquire such information from the environment around it?	Organized, focused inclusion of outside ideas and opinions; broad horizon	Uncoordinated; only internal ideas are pursued further
Client Integration	Role of the client	What is the role of the client on the way from an idea to the launch of a new offering? (Prototype construction or joint tests)	Programs for involving the customer forms of cooperation (prototyping); early inclusion in development	The client has no role; quite late involvement
Self-Reflectiveness	Review process	How consciously and how frequently is the corporate or departmental strategy reviewed? (Self-critiquing)	Regular, not limited to the management level	Unintentional; limited to top management
	Organizational slack	Is there room for rethinking, criticizing, and debating or does the everyday routines not allow this? (Slack time)	Time is available or made available	No time available

Current research tends to subsume these elements of the third degree under the term of 'dynamic capability'. Sammerl reviews several definitions of this term and concludes that all relate to the development and modification (made possible by the right aptitude for learning) of existing resources and competences [33].

The third degree is conceptually translated into the variables of 'receptiveness', 'client integration', and 'self-reflectiveness', with the indicators relating to the sourcing of information, the role of the client, the review process, and the presence of organizational slack.

1.2.7.5 Fourth Degree of Maturity

The fourth degree of maturity is determined by considerations of path dependency or, specifically, the ability to liberate the company from it (Table 1.15). A common trait of the lower degrees of maturity – termed 'dynamic capabilities' – in short is their strong focus on existing resources. When it comes to the ability to innovate, these degrees of maturity will encounter natural limits that can only be crossed if additional competences are added to the equation. The reflective abilities mentioned in the third degree of maturity are important for being able to recognize positive feedback mechanisms, but they are not sufficient to overcome the constraints of path dependency. The additional management competences required here relate to the observers' and consultants' role and to the ability to decide in favor of short-term efficiency or long-term progress.

The fourth degree of maturity is expressed in the variables of 'awareness of path dependency' and 'emancipation from path dependency'. Observation and decision-making represent the indicators relating to these variables.

1.2.7.6 Fifth Degree of Maturity

The elements of the first four degrees of maturity equip companies with substantial (decision-making) abilities, sufficient resources, organizational structures, and the ability to learn and stay receptive to information. Reflection skills allow the anticipation of the risks of path dependency. Finally, decision-making skills can help overcome path dependency or respond professionally to the conundrum of short-term versus long-term considerations. However, none of these elements automatically guarantee a successful launch of novel services, which the present study suggests to be an important condition for service innovation. The fifth degree of maturity alone equips companies with the applied competences that they need to launch novel services on time and with systematic and methodical professionalism in the sense of a good time-to-market approach (Table 1.16).

The fifth degree of maturity is expressed in the variable of 'methodology'. The timing of market entries, financial controlling, and project management tools are the relevant indicators in this respect.

Table 1.15. Fourth degree of maturity – variables, indicators, and key questions. Source: Illustration by the author.

Variable	Indicator	Key Question	Indicator	
			Achieved	Not Achieved
Awareness of Path Dependency	Observation	In which occasions does an external observer come into the equation (institutionalized observation or strategic conferences)?	The role is known and given institutional form, used comprehensively in leadership; methods are available, time is invested; awareness for the significance of the role	The role is not known; there are no methods for this purpose; no management attention and no time
Emancipation from Path Dependency	Decision-Making	To what extent are instruments (performance target systems) designed to anticipate the short *and* long-term implications of decisions?	Distinct urgency, timing, and resources for long-term projects	One-sided outlook; focus on efficiency; no methods or mechanisms in place

Table 1.16. Fifth degree of maturity – variables, indicators, and key questions. Source: Illustration by the author.

Variable	Indicator	Key Question	Criteria
Methodology	Time of market entry	How do you determine when to launch a new offering in the market?	Means for timing the market entry are being used
	Financial controlling	How do you ensure the commercial viability of the new offering? (Financial planning and controlling)	Controlling instruments are available and are being used
	Project management tools	How are ideas turned into market-viable services? (Teams for development and implementation, specifications, testing, etc.)	Implementation with project management methods

1.2.7.7 Summary

The degrees of maturity proposed above can be subsumed in one maturity model (Table 1.17

Table 1.17. Innovative capacities maturity model. Source: Illustration by the author.

Aspects of innovative capacities	Degrees	MD1	MD2	MD3	MD4	MD5
Resources, Competence Degree 1						
Human assets	1					
Brand	1					
Structures (Ambidextrous Organization)						
Organization	2					
Leadership	2					
Culture	2					
Aptitude for Learning						
Receptiveness	3					
Client Integration	3					
Self-Reflectiveness	3					
Path Dependency						
Awareness	4					
Emancipation	4					
Implementation						
Methodology	5					

1.3 Empirical Study

The 'technology focus gauge' and the 'innovative capacities maturity model' are used as empirical tools for measuring the possible link between the service providers' technological affinity and their ability to innovate.

The selection of the service providers chosen for this study is made by means of the technology focus gauge (TFG), which establishes a broad spread of participating companies, TFG 1 to 4, for the comparison. The choice of experts at these companies takes their experience and knowledge into account in that they can be expected to provide meaningful expert answers for the questions developed from the maturity model. The analysis of their responses was used to allocate a level of maturity to each company and to allow a conclusion about the possible correlation with their technology focus.

1.3.1 Selection of Research Objects

The selection of the participants for the empirical study was made on the basis of the above definition of service providers and their technology focus as assessed by means of the gauging tool described before. The first factor, i.e. service provision, was simple to establish, whereas the actual challenge lay in the rating of the companies' technology focus: this had to be produced before establishing contact with the target participants. The eventual selection was therefore the product of an analysis of their service offerings and a general company analysis, aimed at ensuring diversity in the sample in terms of company sizes, offerings, and customer groups.

Table 1.18 introduces all research objects with the allocated technology levels.

Table 1.18. The research objects in brief. Source: Author's data.

Company	Sector	Size (headcount)	TGF*
Kresta, Schnider & Partners	Consulting	14	1
Hotelplan Suisse	Tour operator	1,500	2
Schweizerische Mobiliar	Insurance	3,500	2
Entris Banking	Banking services	315	3
StudiMedia	Academic marketing	7	3
Numcom	Software	40	4
Swisscom IT Services	IT services	2,800	4
IBM (Schweiz)	IT services	3,250	4

* TFG: Technology focus gauge. 1=weak, 4=strong

1.3.2 Selection of Experts

The purpose of the experts' selection lay in the identification of a person whose sophisticated insights, extensive experience, and awareness of the general relations between the market's needs and their company's innovative abilities enabled them to produce qualified judgments. The ideal expert possesses special access to information about groups of people or decision-making processes and has responsibility for certain solutions within the business [54].

The following criteria were applied in the selection of the experts. They were not cumulatively exclusive criteria, but treated as general guidelines:
- long-standing experience at the company in question
- profound knowledge of service development
- executive function at the company
- willingness to participate and availability for the expert interviews

1.3.3 Empirical Tools

The eventual selection of eight companies with different degrees of technological affinity was investigated by means of structured expert interviews.

The development of the innovative capacities maturity model has defined all variables, indicators, and relevant questions and measures to answer the study questions. This also provided the basis for the collection and processing of the data. The five maturity levels translated into five thematic parts of the empirical tool, which were expressed in the sections of the interview guidelines, the analysis of the data, and the interpretation of the results. The interview guidelines were also subjected to a pretest.

1.3.4 Analysis

All eight interviews were recorded digitally and transcribed in full. The participants' responses were allocated to the defined thematic parts or their subdivisions. The aim of this approach was to establish the maturity of each company's innovative capacities, revealing any potential correlation when contrasted with their defined technology focus.

1.3.4.1 Assessment of the Maturity Model

The previous sections have defined the eight participating companies' achievement of the five maturity levels. The results are visualized in Figure 1.2.

The narrow sample notwithstanding, the results seem significant. There is no apparent correlation between technology focus and innovative capacities as described in this study.

Although KSP is perceived as the company with the lowest technological affinity, it can be accredited with maximum innovative capacities. The two companies with the strongest technology focus, by contrast, are awarded maturity level 2. This indistinct picture is reinforced by the three companies attributed with third-degree technology focus, but innovative capacities graded at levels 1, 3, and 5, respectively. Studimedia, Hotelplan, and Mobiliar are close to a line that would imply a linear correlation between technology focus and innovative capacities, but the other five surveyed companies deviate from that line by a considerable margin.

1.3.4.2 Points System Replacing the Maturity Model

When allocating the research objects to the right maturity level for their innovative capacities, all higher levels are excluded after the first level of maturity that is not attained. This can represent a disadvantage for the company in question, since certain qualities are forcibly excluded from the final assessment. The question of whether a

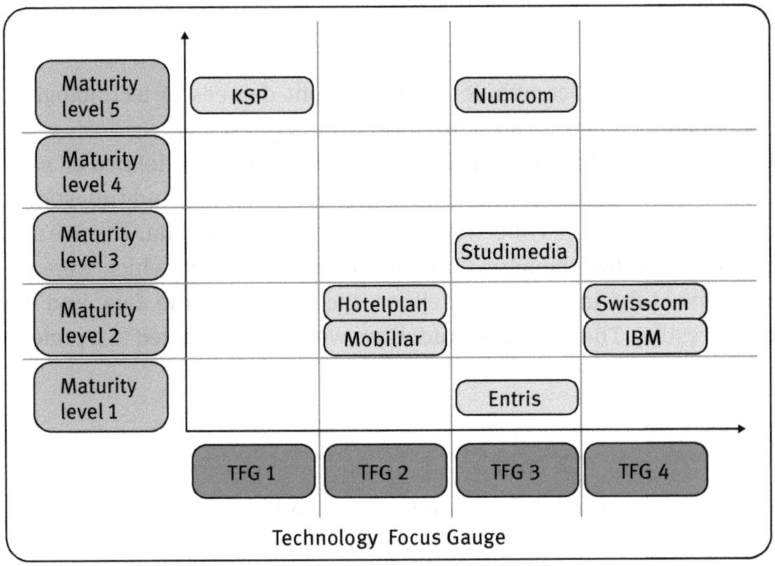

Fig. 1.2. Results of the maturity model

maturity model is the suitable choice for this case in view of its essential exclusive structure should therefore be revisited. In particular, the following aspects might have a negative effect on the final assessment:

- sequence of the variables in each maturity level
- validity of the questions for the variable's indicators
- quality of the questions asked in the interview
- interpretation of the responses

Figure 1.3 contrasts the maturity model employed here with a points system combining all points for each applicable level of maturity. Even with this calculation, no correlation between innovative maturity and technology focus could be ascertained.

1.4 Conclusions and Discussion

Our study ends with a discussion of the findings and conclusions as well as a critical assessment of the investigation.

1.4.1 Conclusions

The results are unequivocal for both the maturity model and the points system: There is no evident correlation between a company's technology focus and its ability to in-

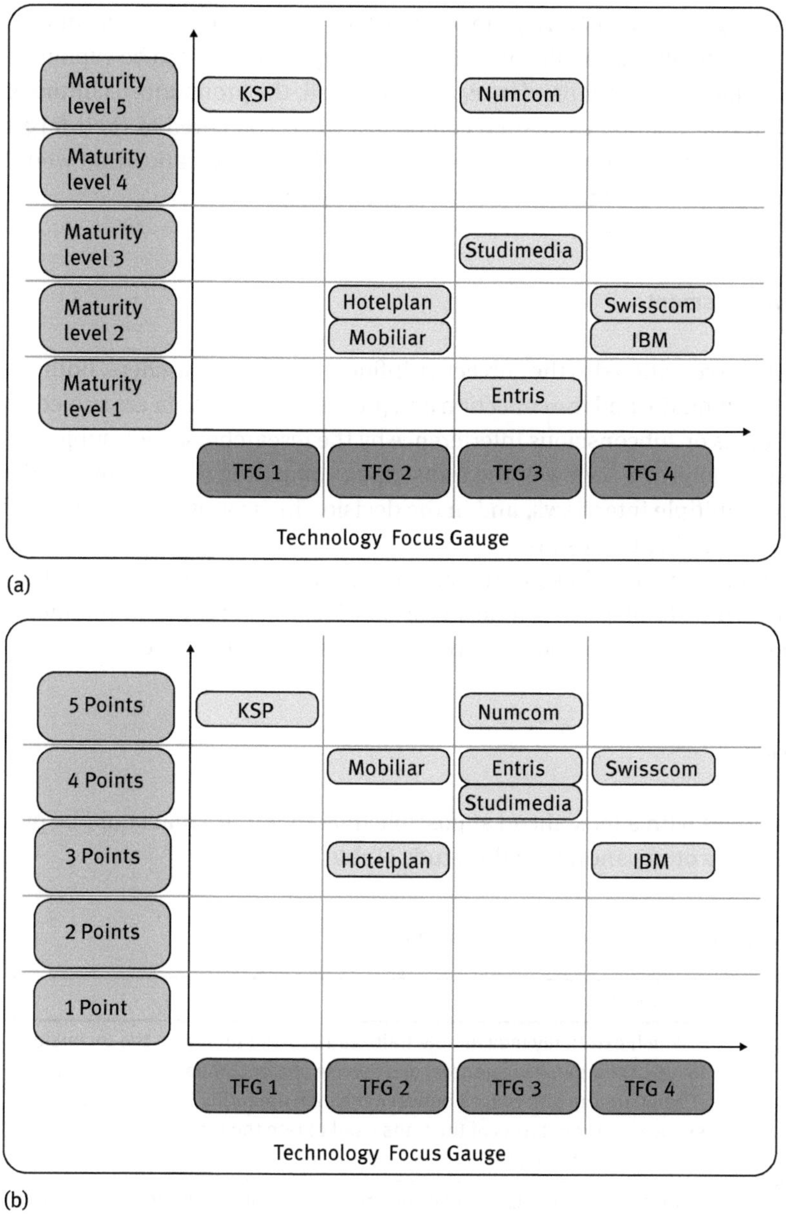

Fig. 1.3. Results of the maturity model (a) vs. points system (b)

novate. It is striking that the maturity model proposed here awards greater innovative capacities in particular to smaller and younger enterprises. This can be explained with their greater flexibility, independence, and, above all, the pronounced culture of entrepreneurship that dominates at such companies when compared to their larger peers. Naturally, this conclusion should only be regarded as an indication that would need closer scrutiny in view of the small scale of the inquiry.

1.4.2 Critical Assessment

The first factor to be considered is the subjective influence of the researchers. Both the design of the questionnaire and the collection and processing of the data are subject to the risk of conscious or subconscious interference by the researchers. Such subjective biases are most obvious in a review of the transcripts, comparing the phrasing of the questions across multiple interviews, and in the decision in favor of a certain level of maturity after interpreting the results.

The second factor again relates to the narrow scope of the inquiry. An analysis of the link between technology and innovation that has the ambition of producing statistically meaningful results is not possible under these circumstances.

1.4.3 Discussion

This paper concludes with a look ahead at possible lines of inquiry that could be pursued by following up or expanding on this study (Table 1.19).

Table 1.19. Follow-up questions for the current comparative study.

Issues	Questions
Technology focus	– Which externally measurable traits can be used to assess the technology focus of companies in general and service sector companies in particular? – What are the differences between the manufacturing industry and the service sector in terms of the traits used to rate the technology focus of industry companies?
Innovative capacities	– Are there similar approaches for measuring innovative capacities and how do they differ from the maturity model proposed here?
Client's integration	– Which options are there for promoting innovation by integrating clients in a mass market?

Bibliography

[1] Europäische Komission: Übersicht Industriezweige. Online: http://ec.europa.eu/enterprise/sectors/index_de.htm (Stand: 4. April 2011), 2011.

[2] Tschirky, Hugo: Technologie-Management: Schliessung der Lücke zwischen Management-Theorie und Technologie-Realität. In: Tschirky, Hugo; Koruna, Stefan (Hrsg.): Technologie-Management – Idee und Praxis. Zürich, Orell Füssli Verlag. S. 1–32, 1998b.

[3] Maler, Rudolf: Grundlagen der Dienstleistungsproduktion. Berlin, Springer Verlag, 1997.

[4] Kleinaltenkamp, Michael: Begriffsabgrenzungen und Erscheinungsformen von Dienstleistungen. In: Bruhn, Manfred; Meffert, Heribert (Hrsg.): Handbuch Dienstleistungsmanagement. Wiesbaden, Gabler. S. 27–50, 2001.

[5] Corsten, Hans; Gösinger, Ralf: Dienstleistungsmanagment. Oldenburg, Oldenburg Verlag.

[6] Perl, Elke: Grundlagen des Innovations- und Technologiemanagements. In: Strebel, Heinz (Hrsg.): Innovations- und Technologiemanagement. Wien, WUV Universitätsverlag. S. 15–48, 2003.

[7] Radnitzky, Gerard: Wissenschaftstheorie, Methodologie. In: Seiffert, Helmut; Radnitzky, Gerard (Hrsg.): Handlexikon zur WIssenschaftstheorie. München, Ehrenwirth Verlag GmbH. S. 463–472, 1989.

[8] Ropohl, Günter: Allgemeine Technologie: eine Systemtheorie der Technik. (3., überarbeitete Auflage) Karlsruhe, Universitätsverlag Karlsruhe, 2009.

[9] Tschirky, Hugo: Konzept und Aufgaben des integrierten Technologie-Managements. In: Tschirky, Hugo; Koruna, Stefan (Hrsg.): Technologie-Management – Idee und Praxis. Zürich, Orell Füssli Verlag. S. 193–394, 1998a.

[10] Brodbeck, Harald: Strategische Entscheidungen im Technologie-Management. Relevanz und Ausgestaltung in der unternehmerischen Praxis. Zürich, Orell Füssli Verlag, 1999.

[11] Arvanitis, Spyros; Hollenstein, Heinz; Marmet, David: Internationale Wettbewerbsfähigkeit. Wo steht die Schweiz? Zürich, vdf Hochschulverlag, 2005.

[12] Seeger, Heike: Ex-Post-Bewertung der Technologie- und Gründerzentren durch die erfolgreich ausgezogenen Unternehmen und Analyse der einzel- und regionalwirtschaftlichen Effekte. Münster, Lit Verlag, 1997.

[13] Thudium, Thomas: Technologieorientiertes strategisches Marketing: Die Entwicklung eines neuen Bezugsrahmens zur Generierung von Marketingstrategien für technologieorientierte Unternehmen. Wiesbaden, Gabler, 2005.

[14] Kulicke, Marianne: Chancen und Risiken junger Technologieunternehmen: Ergebnisse des Modellversuchs "Förderung technologieorientierter Unternehmensgründungen". Heidelberg, Physica-Verlag, 1993.

[15] Bolz, Alexander E.: Innovation, Kooperation und Erfolg junger Technologieunternehmungen: Konzept – Panelstudie – Gestaltungsempfehlungen. In: Corsten, Hans; Reiss, Michael; Steinle, Claus; Zelewski, Stephan (Hrsg.): Information – Organisation – Produktion. Wiesbaden, Gabler, 2008.

[16] Gelbmann, Ulrike; Vorbach, Stefan: Strategisches Innovations- und Technologiemanagement. In: Strebel, Heinz (Hrsg.): Innovations- und Technologiemanagement. Wien, WUV Universitätsverlag, 2003.

[17] Tschirky, Hugo: Technologie-Management: Schliessung der Lücke zwischen Management-Theorie und Technologie-Realität. In: Tschirky, Hugo; Koruna, Stefan (Hrsg.): Technologie-Management – Idee und Praxis. Zürich, Orell Füssli Verlag. S. 1–32, 1998b.

[18] Scherngell, Thomas: Interregionale Wissensspillovers in der europäischen High-Tech Industrie. Eine empirische Analyse. (1. Auflage) Wiesbaden, Deutscher Universitätsverlag/GWV Fachverlage GmbH, 2007.

[19] Hatzichronoglou, Thomas: Revision of the High-Technology Sector and Product Classification. OECD Science, Technology and Industry Working Papers, No. 1997/02. doi: 10.1787/134337307632, 1997. Online: http://www.oecd-ilibrary.org/science-and-technology/revision-of-the-high-technology-sector-and-product-classification_134337307632. Stand: 2. April 2011.

[20] OECD: OECD Science, Technology and Industry Scoreboard 2007. OECD Publishing. doi: 10.1787/sti_scoreboard-2007-en, 2007. Online: http://www.oecd-ilibrary.org/science-and-technology/oecd-science-technology-and-industry-scoreboard-2007_sti_scoreboard-2007-en. Stand: 2. April 2011.

[21] Lessat, Vera; Eckerle, Tobias H; Licht, Georg; Nerlinger, Eric; Steiger, Max: Beteiligungskapital und technologieorientierte Unternehmensgründung. Wiesbaden, Gabler, 1999.

[22] Hübner, Heinz: Integratives Innovationsmanagement. Nachhaltigkeit als Herausforderung für ganzheitliche Erneuerungsprozesse, Berlin, Erich Schmidt, 2002.

[23] Strebel, Heinz: Innovations- und Technologiemanagement. (2. Auflage) Wien, Facultas, 2007.

[24] Bergmann, Gustav; Daub, Jürgen: Systemisches Innovations- und Kompetenzmanagement. Grundlagen – Prozesse – Perspektiven. (2. Auflage) Wiesbaden, Gabler, 2008.

[25] Haber, Tobias E.: Resistenz gegenüber Innovationen. (1. Auflage) Wiesbaden, Gabler, 2008.

[26] Schmalen, Helmut; Pechtl, Hans: Die Rolle der Innovationseigenschaften als Determinanten im Adoptionsverhalten, in: Schmalenbachs Zeitschrift für betriebswirtschaftliche Forschung, Vol. 48, 1996, S. 816–836, 1996.

[27] Hauschildt, Jürgen; Salomo, Sören: Innovationsmanagement. (5.Auflage) München, Vahlen, 2011.

[28] Schniering, Nina: Industrielle Dienstleistungsinnovation. Eine praxisorientierte Analyse der Determinanten und Erfolgswirkungen. Berlin, Lit Verlag Dr. W. Hopf, 2009.

[29] Busse, Daniel: Innovationsmanagement industrieller Dienstleistungen. Theoretische Grundlagen und praktische Gestaltungsmöglichkeiten. (1. Auflage) Wiesbaden, Gabler, 2005.

[30] Ili, Serhan: Open Innovation umsetzen. Prozesse, Methoden, Systeme, Kultur. (1. Auflage) Düsseldorf, Symbosion Publishing GmbH, 2010.

[31] Chesbrough, Henry W.: Open Innovation: the new imperative for creating and profiting from technology. Boston, Harvard Business School Press, 2003.

[32] Reichwald, Ralf; Piller, Frank: Interaktive Wertschöpfung. Open Innovation, Individualisierung und neue Formen der Arbeitsteilung. (2. Auflage) Wiesbaden, Gabler, 2009.

[33] Sammerl, Nadine: Innovationsfähigkeit und nachhaltiger Wettbewerbsvorteil. Messungen – Determinanten – Wirkungen. (1. Auflage) Wiesbaden, Gabler, 2006.

[34] Burr, Wolfgang; Stephan, Michael: Dienstleistungsmanagement. Innovative Wertschöpfungskonzepte für Dienstleistungsunternehmen. Stuttgart, Kohlhammer, 2006.

[35] Kühnl, Christina: Erfolgsfaktoren im Innovationsmanagement. Eine empirische Betrachtung aus Hersteller- und Anwendersicht. (1. Auflage) Wiesbaden, Gabler, 2010.

[36] Hamel, Gary; Prahalad, C. K.: Competing for the future. Boston, Harvard Business School Press, 1994.

[37] Stein, Volker; Klein, Torsten: Organizational Slack als Dynamisierungsquelle organisationaler Kompetenzen. In: Stephan, Michael; Kerber, Wolfgang (Hrsg.): Jahrbuch Strategisches Kompetenzmanagement. Band 4: "Ambidextrie": Der unternehmerische Drahtseilakt zwischen Ressourcenexploration und -exploitation. München, Mering, Rainer Hampp, 2010.

[38] Freiling, Jörg; Gersch, Martin; Goeke, Christian: Eine "Competence-based Theory of the Firm" als marktprozesstheoretischer Ansatz. In: Schreyögg, Georg; Conrad, Peter (Hrsg.): Management von Kompetenz. Managementforschung 16. (1. Auflage) Wiesbaden, Gabler, 2006.

[39] Gomez, Peter; Raisch, Sebastian; Rigall, Juan: Die Formel für profitables Wachstum. Harvard Business Manager. Sonderdruck aus dem Juli-Heft 2007. Online: http://www.droege-international.com/fileadmin/downloads/070701_HBM_Formel-fuer-profitables-Wachstum.pdf (Stand: 17.03.2011).

[40] Schreyögg, Georg: Unternehmenskultur zwischen Stabilität und Wandel. In: Lange, Hellmuth; Manske, Fred (Hrsg.): Kultur im Veränderungsprozess. Baden-Baden, Nomos. S. 23–36, 2004.

[41] Kühl, Stefan; Schnelle, Wolfgang: Laterales Führen. Wenn Hierarchie nur begrenzt zur Verfügung steht. In: Aderhold, Jens; Meyer, Matthias; Wetzel, Ralf (Hrsg.): Modernes Netzwerkmanagement. (1. Auflage) Wiesbaden, Gabler, 2005.

[42] Cohen, Wesley M.; Levinthal, Daniel A.: Absorptive capacity: A New Perspective on Learning and Innovation. In: Administrative Science Quarterly, Volume 35, Issue 1, S. 128–152, 1990. Online: http://faculty.fuqua.duke.edu/~charlesw/s591/Bocconi-Duke/Papers/C10/CohenLevinthalASQ.pdf (Stand: 08.04.2011).

[43] Schreyögg, Georg: Absorptive Capacity – Schlüsselfaktor für die Innovationsfähigkeit 1. Innovationsquelle externes Wissen – Die Rolle der Absorptive Capacity. Innovationsforum der Daimler und Benz Stiftung, 2010. Online: http://www.daimler-benz-stiftung.de/cms/uploads/images/service/downloads/Essay_Absorptive%20Capacity_2.pdf (Stand: 17.03.2011).

[44] Steinmann, Horst; Schreyögg, Georg: Management. Grundlagen der Unternehmensführung. Konzepte – Funktionen – Fallstudien. (6. Auflage) Wiesbaden, Gabler, 2005.

[45] March, James G.; Olsen, Johan P.: Ambiguity and choice in organizations. (2nd edn.) Bergen, Universitetsforlaget, 1979.

[46] Argyris, Chris; Schön, Donald A.: Organizational Learning II. Theory, Method, and Practice. Addison-Wesley Publishing Company, 1996.

[47] Wimmer, Rudolf: Vorausschauende Selbsterneuerung – Wie sich Organisationen mit lebenswichtigen Irritationen versorgen. In: Hinterhuber, Hans H.; Stahl, Heinz K. (Hrsg.) Fallen die Unternehmensgrenzen? (1. Auflage) Renningen, Expert Verlag, 2001. Online: http://www.osb-i.de/fileadmin/user_upload/Publikationen/Wimmer_Vorausschauende_Selbsterneuerung.pdf (Stand: 23.03.2011).

[48] Bruhn, Manfred; Stauss, Bernd: Kundenintegration im Dienstleistungsmanagement – Eine Einführung. In: Bruhn, Manfred; Stauss, Bernd (Hrsg.) Kundenintegration. Forum Dienstleistungsmanagement. (1. Auflage) Wiesbaden, Gabler, 2009.

[49] Vesshoff, Julia; Freiling, Jörg: Kundenintegration im Innovationsprozess. Eine kompetenztheoretische Analyse. In: Bruhn, Manfred; Stauss, Bernd (Hrsg.) Kundenintegration. Forum Dienstleistungsmanagement. (1. Auflage) Wiesbaden, Gabler, 2009.

[50] Dievernich, Frank E. P.: Pfadabhängigkeit im Management. Stuttgart, Kohlhammer, 2007.

[51] Sydow, Jörg; Schreyögg, Georg; Kochen, Joch: Organizational Paths: Path Dependency and Beyond. 21st EGOS Colloquium, Berlin, 2005. Online: http://www.wiwiss.fu-berlin.de/forschung/pfadkolleg/downloads/organizational_paths.pdf (Stand: 09.04.2011).

[52] Stern, Thomas; Jaberg, Helmut: Erfolgreiches Innovationsmanagement. (4. Auflage) Wiesbaden, Gabler, 2010.

[53] Gläser, Jochen; Laudel, Grit: Experteninterviews und qualitative Inhaltsanalyse. (3. Auflage) Wiesbaden, GWV Fachverlage GmbH, 2009.

[54] Meuser, Michael; Nagel, Ulrike: ExpertInneninterviews – vielfach erprobt, wenig bedacht: ein Beitrag zur qualitativen Methodendiskussion, in: Bogner, A.; Littig, B.; Ment, W. (Hrsg.): Das Experteninterview. Theorie, Methode, Anwendung. (2. Auflage) Opladen, Westdeutscher Verlag, 2005.

Felix Adamu Nandonde*, Galinoma Lubawa, and Pamela John Liana

2 Uptake Of Market 'Induced Innovation' by Upstream Actors in Tanzania

Abstract: This chapter investigates the uptake of innovation induced by downstream actors by sunflowers SMEs in Tanzania with the use of an interview-based qualitative study of data collected in Dodoma, Tanzania. This study reveals that SMEs pay less attention to organizational innovation. However, they are much more focused on production innovation in the interest of servicing markets. Furthermore, the study reveals that a major factor that influences the uptake of innovation from downstream are not final consumers, but rather government agencies. For instance, most SMEs involved in the study search for funds to buy small oil refinery machines after government bans of the selling of raw oil. This suggests that processors still think consumers will absorb everything because there is lack of supply and demand is high.

2.1 Introduction

With the rapid rise of modern food retailing in developing countries, food suppliers experience spectacular changes in the value chain. One of the changes is the emergence of the retailers and consumers as the drivers of the value chain. This means that suppliers have to produce what the market wants, such as packaging materials, style, quality and the kind of information shared on the label. In this work, this phenomenon is referred to as 'market induced innovation'. This chapter responds to an appeal by Beuleans et al. for more research on how SMEs in agri-food businesses respond to and manage value chain innovation induced from downstream [1]. To achieve this objective the proposed chapter uses interviews with sunflower cooking oil processors in Tanzania to investigate how SMEs respond to innovation uptake from downstream.

Demand for cooking oil in Tanzania is estimated to be 250,000 metric tons per year in 2011 and is expected to grow between 5 and 6 percent annually [2]. Only 40 per cent is domestically produced [3]. It is estimated that the country imports 300,000 tons of cooking oil annually [4] which amounted to US$ 120 million [5]. Importation combined with domestic production exceeds the estimated domestic demand by 50,000 metric tons, indicating that perhaps some may be exported to neighboring countries. Importation of edible oil threatens the growth of locally processed oil. On the other

***Corresponding Author: Felix Adamu Nandonde:** Aalborg University, Denmark, fan@business.aau.dk, nandonde@gmail.com
Galinoma Lubawa: The Institute of Rural Development Planning, Tanzania
Pamela John Liana: The Open University of Tanzania

hand, failure of local oil processors may be due to lack of innovation to meet market demand.

The major producers of cooking oil in Tanzania are Mount Meru in Arusha, with an estimated capacity of 50,000 metric tonnes per year, and Murzah Oil, with its factories in Dar-Es-Salaam.

Kuada states that there is lack of study of innovation in Africa [6]. Some of the few studies that have been conducted in Tanzania have focused on innovation at SMEs and in the manufacturing sector. De Bruijin and Mahemba conducted a study on SMEs innovation in Tanzania and identified that active relations among actors influence innovation [7]. Further, their study reveals that there is no relationship between the membership of business organizations and the innovative behavior of the SMEs. Another study by Musambala in Tanzania found that rural sunflower oil processors learn more from their competitors and that technology learning transfers take place based on the availability of appropriate technologies [8]. Despite great efforts to investigate innovation and SMEs in Tanzania there is less focus on the uptake of the innovation induced by downstream actors.

It is well known that innovation can be in the form of technology, market, or process, and can be accelerated internally or externally. We argue that manufacturers pushed product to the retailers without having the knowledge of the demand. The current chapter argues that the emergence of retailers as the driver and gatekeepers of different food items means that the supplied product has to meet market demand. This point highlights the importance of issues related to innovation.

It is important to understand the influence of different actors in the value chain with focus on upstream and downstream. Therefore, this chapter intends to fill that gap by investigating the uptake of downstream innovation by SMEs in Tanzania.

2.1.1 The concept of innovation

The emergence of consumers as the major drivers and influencing partners in the value chain introduced change from supply-push to demand-pull. With supply-push innovation the suppliers determine what to produce and when to deliver. In demand-pull consumers determine what to produce and when to produce. Therefore, demand-pull innovation requires that all actors in the production network have to work together to meet the market demand. The above market situation implies that innovation is no longer linear. There are many actors, such as retailers, consumers, regulators and suppliers having direct impact on the innovation process.

What is innovation? The concept of innovation is essential for the firm's success in the market place. Despite the concept being one of the reasons for the firm's success, there is no consensus of what innovation is [9]. In the food industry innovation is very important too, however the sector is one that has a low level of innovation compared with other industries.

Beuleans et al. define innovation as technologies invested or adapted in the focal market [10]. This definition intends to take technology as the source of all innovation. However, Omar argues technology alone could not be a source of innovation [11].

Grunert at al. define innovation as a process towards the development of a new product or service in which an integrated analysis and understanding of the users' wants, needs and preference formation play a key role [12]. This definition being inclusive of all actors in the value chain, especially downstream players such as customers, it has been criticized for being too broad by including customers and end users in consumer-oriented innovations that affect multiple actors of value chain [13]. On the other hand, retailers are reducing the number of actors in the value chain, which implies they like to work with processors or growers directly due to market competition. Given this process, it is obvious that some processors would be removed in some commodities' value chains in the interest of retailers. The emergence of private brands further reduces the value chain.

Schumpeter referred to innovation as the creation of new combinations [14, 15]. Schumpeter's definition of innovation is based on the assumption that innovation can take place at a stage of production for the improvement of delivery of product regardless of the level of technology used. One of the major problems that faces Africa' food sector is the availability of high quality products all year around at all locations [16]. Therefore, important innovations are not taking place in the laboratory alone – some innovations must take place at the market level, such as changes in selling and procurement policies. This study adopts Schumpeter's definition because it is associated with either breakthrough or incremental innovation that occurs at any stage, such as production, process, or organization.

2.2 Literature review

2.2.1 Innovation in the value chain

Figure 2.1 shows that there are several elements of value chain that influence decisions of the processor regarding innovation. External factors influence innovation of the downstream actors, such as retailers and suppliers, in many ways, such as through policies, regulations and laws. For example, governmental agencies in a particular country can ban the use of certain ingredients, but the same material or input can be used in the neighboring countries. Furthermore, because suppliers and retailers don't have machinery to control pirating, the higher the level of enforcement the more the chances actors in the value chain have to be innovative.

Consumers are major players in the innovation of the actors in the upstream and downstream. However, the decisions of consumers are influenced in some situations by the countries' policies and information they receive from non-governmental organizations (NGOs) and government agencies.

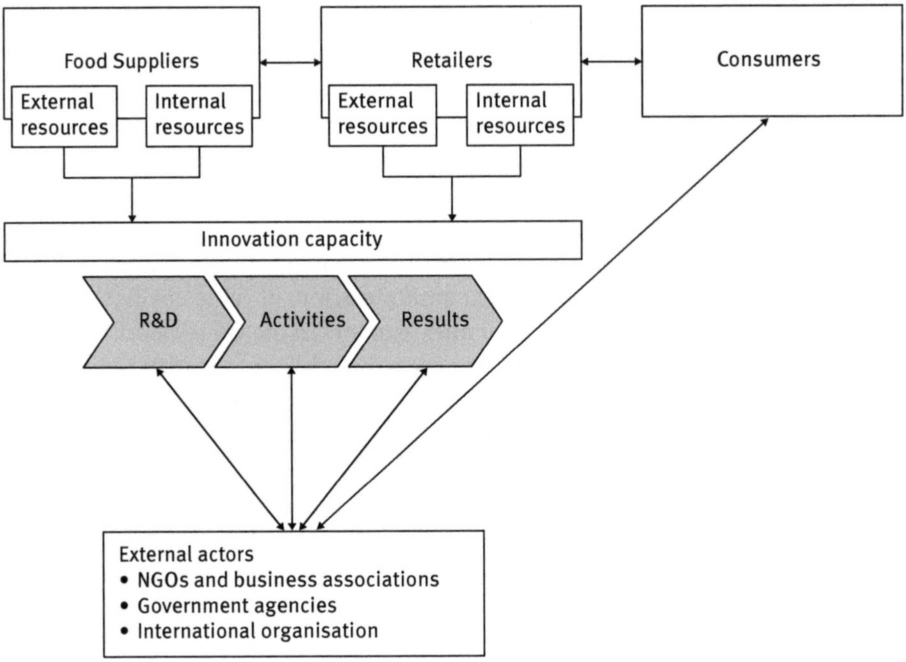

Fig. 2.1. Innovation capacity at company and value chain. Source: Modified from [17–19]

Internal resources refer to the research and development (R&D) structure and a vast number of firms' characteristics such as size, financial structure, qualified staff, experience of the manager and openness to new ideas, all of which influence innovation processes ([20, 21]).

External resources refer to the firms' strategic environment and include the potential for business-to-business relationships, available infrastructure for collaboration and networking, and access to support from research providers and government [22]. Here the focus is on the environment in which firms operate its impacts on the firms' innovations. Gellynck et al. argue however that researchers have largely overlooked the ubiquitous influence of the institutional environment and how interorganizational relationships, such as marketing channels, are embedded in the larger social context [23]. The firm's external network support to foster firms' innovation is very important in considering how the firms establish relationships with other actors in the network. Figure 2.1 shows that there is direct contact between retailers and consumers, however decisions of retailers are greatly influenced by external factors such as NGOs, government agencies and international organizations. Gellynck et al. argue that external drivers of innovation are forces caused by regulation, changes in the food sector and social pressures [23]. So for any innovation to be meaningful from either a firm level or an industrial level, it has to get support from the external actors who have direct influence on the consumer's purchases. Scozzi et al. found that government regulation influences network structure in India [24]. As a result, government policies can influence trust and power between actors in the network which may influence the sharing of information among actors in the network.

Levels of innovation

Innovation in the food industry can be categorized into three levels, which are process, product and organization innovations. Empirical evidence indicates that innovation in the food industry is at the level of process and product innovation [25, 26]. Furthermore, research shows that SMEs in the food industry focus much more on process and product innovations, and ignore organizational innovation [27].

2.2.1.1 Product innovation

Product innovation can be an old product in a new market, or a new product with new features, or an old product that has added new features. In the food industry firms can introduce new innovation with packaging or change of brand color or style and shape while the ingredients can be the same or slightly different. In general, product innovation can range from radical to incremental innovation.

1. Radical innovation causes marketing and technological discontinuities at both the macro- and micro-level.
2. Incremental innovation occurs at the micro level and causes either a marketing or technological discontinuity but not both.

Both radical innovation and incremental innovation have the advantages of adding to the economic growth of a country and to the performance of the individual. However, in the food industry R&D occurs rarely at the SMEs due to the resources required and the risk associated with it.

2.2.1.2 Process innovation

The process of innovation involves creation, design, production, first use and diffusions of new products, their processes, services and or systems [28]. At this stage innovation heavily depends on technology and human skills. The firm size can be a major source of innovation, with its resource capabilities of employing skilled staff and introducing appropriate technology to support innovation. However, technology in not necessary, depending on the technology, for success, but diffusions is very important. Therefore at the firm level there must be link between its innovation efforts and marketing innovation. This will be explained in Section 2.2.1.3.

2.2.1.3 Organizational innovation

With the increase of competition in the food value chain, actors in the network have to deliver products on time and at the right price and be more effective. Organizational innovation involves changes introduced by actors that facilitate activities of the other members in the network that focus on the assurance of speed and the availability of product to consumers. Organizational innovation may include implementation of new management tools, changes in staff policy and changes in purchasing and sales policies of the firm. Previous studies in developed countries have shown that actors in the food industry rarely take into consideration organizational innovations [29, 30]. However, organizational innovation might facilitate other types of innovation, such as product and process [31], and could contribute to the performance and effectiveness of the individual firms and the chain networks they operate in [32].

2.2.2 Innovation capacity at the firm level

Figure 2.1 shows that innovation capacity at the firm level is the ability of the firm to develop new products or services throughout all innovation processes, which consists of three steps: efforts, activities and results. However, this capability is determined by the power and flexibility to choose partners among actors in the network [33, 34]. To reduce the cost of changing partners, it is very important for actors in the value chain to identify what each actor would bring in. Further, the focus has to be on human behavior instead of technologies or working capital. Ghauri and Kemp argue that resources are heterogeneous among firms, so to understand how firms may establish collaboration in the chain it's important to differentiate between accessed resources and embedded resources [35]. This may affect the firm's commitments and trust, and opportunistic behavior may occur.

Innovation efforts can be categorized as non-structured human and structured resources [36]. The focus here is to see how human knowledge and skills are used in the organization to develop and introduce new ideas to capture market demand. However, knowledge and skills alone could not have an impact without allocation of resources to finance new inventions such as attending training and trade shows.

Innovation result is the stage at which firms expected to benefit from the efforts they have injected into the innovation at the level of process, product, organizational and marketing strategies. However, to benefit from the sum of the resources invested the firm has to have capacity to not only capture market demand but also to fight against imitations and pirates. Further, government agencies have a great role in influencing entrepreneurs to invest their scarce resources through enforcement of the laws and observing the rule of laws. Therefore, in some industries firms' behavior towards innovation can be influenced by the external environmental and the context in which it operates.

2.2.3 Innovation in the food industry driven by modern retailers

The agri-food industry is generally considered as the low tech sector [37, 38]. However, studies conducted on innovation in the food industry indicate that retailers in developed countries are more innovative due to the increase of their power and introduction of their own brand. In a study conducted in the UK, Omar found that UK retailers are keen on innovation and the introduction of different quality standards as well as employing food technologists since the emergence of their own brand [39]. Nevertheless,Kottila found that in Sweden food manufacturers developed products in-house for consumers and not by working with them inside or outside the supply chain [40]. This indicates lack of involvement of external actors and in particular downstream actors on food innovation processes.

Establishing interorganizational relationships such as alliances, partnerships, collaborations and joint ventures is of increasing important in today's highly competitive market ([41, 42]). Modern food retailing is growing in Tanzania and is very dynamic with a focus on capturing consumers' demands. This creates many pressures on retailers to choose the partners that can enable them to meet consumers' needs in an efficient and effective manner. One of the very important dimensions in interorganizational relationship is the exchange of information. Reardon and Weatherspoon argue that partners in supply chains within a food sector can access new markets and better product design via e-collaboration [43]. Kimeme et al. proved that ICT enables suppliers in Tanzania to implement consumer-driven demand [44].

One pillar that controls the interorganizational relationship is trust in sharing information. The retailer is in contact with many consumers in a day. This gives retailers power to be a major sources of information in the food value chain. However, sharing information among partners that may support innovation depends on the trust that exists between retailers and suppliers.

Previous studies in innovation focused on upstream innovation with interest of understanding buyer-retailer relationships and supply chain structure after the emergence of modern food retailing in developing economies. Dabas et al. found in India government policies and regulation influenced innovation in upstream and supply chain structures [45]. These policies, such as retail price setting which India implements, can limit innovation. Other studies conducted excluded retailers [46] but included buyers like wholesalers and distributors. RLDC found that knowledge-sharing among actors in the organic food industry is very low compared with the information they possess, which limit their opportunities to innovative [47].

Kuada and Gellynck et al. argue that in a relationship with a high level of trust, conflict would be resolved in an early phase and in a way that satisfies both partners [48, 49]. One of the conflicts that emerged since the rise of modern food retailing in Africa is on the retailers' payment policy, a power imbalance whereby retailers punish suppliers if supplied items were not sold for the cost of shelves space [50].

2.2.3.1 Implication for Tanzania

The food industry which is dominated with SMEs is growing, however, it is faced with a number of challenges, such lack of finance. Despite those challenges indigenous entrepreneurs are investing in value addition and upgrading. However, investment in R&D in the food sector is carried out at the university level and other research institutions with the support from donors, government agencies and international organizations.

Previous studies conducted in Tanzania found SMEs in garments and wood markers are accessing information from different sources such as radio, newspaper, and social media to support them in adapting to changes in the business environment ([51]). However, studies conducted in the country have not paid attention to how actors in the value chain exchange information to support their innovativeness. Ndyetabula appeals for a study that takes into consideration the external environment of agri-food firms in Tanzania in order to understand its impact on innovation and entrepreneurship [52]. The current project induced food innovation in a value chain network within external business environments.

Tanzania and Africa in general have experienced massive changes in its urban food landscape. This has been accelerated with the increase of income and swelling of the middle class.

2.3 Methodology

2.3.1 Data collection

A qualitative method was employed for this study and interviews were used for data collection. Data were collected from SME sunflower processors in Dodoma, Tanzania. Sunflower oil processors were selected because sunflower is one of the oil seeds that has reported a great improvement in the country compared with other seeds such as palm oil, ground nuts, cotton seeds and sesame. A recent study found that innovation spreads very fast in the sunflower processing sector in Tanzania compared with other oil seed commodities [53]. Sunflowers are grown in different part of the country, such as Dodoma and Singida, however Dodoma was selected because the sunflower processing sector is highly advanced due to the number of initiatives taken at the macro, meso and micro level. Furthermore, the Dodoma region is major producer of sunflowers in the country, and is estimated to produce 22 per cent of the country's production [54]. These reasons meant selection of Dodoma as the area for studying the innovation uptake of SMEs is viable compared with other parts of the country.

Participants included in the study are owners, managers, accountants and directors. Table 2.1 shows seven participants that were recruited for the study, all from Dodoma as our previous argument for selection suggests. One owner of modern retailing firms is found in the same city. SME firms were selected based on previous experience of the researchers working with them in various training and research programmes. Therefore, approaching and contacting the processors was based on snowballing and previous experience. Further, availability of their food produce in various food retail stores was used by researchers to decide which firms to contact.

Interviews were conducted at SMEs' premises after an agreement with researchers when and at what time to visit them. Data were collected with two researchers with a tape recorder or camera after getting the acceptance of interviewee. One interviewer was tasked with controlling the interview process and asking questions while the second interviewer took notes. A semi-structured interview was used to enable researchers to have similar questions for all participants in the study. Semi-structured questions were developed from various literature (e.g. [55, 56]). The original semi-structured interview was developed in English and later translated in Kiswahili which is a common language in Tanzania. A Kiswahili checklist was used for the interview and all interviews were collected using the Kiswahili language. On average each interview took 40 minutes. All recorded information was transcribed for further analysis. Qualitative data were manually handled.

Table 2.1. Description of respondents' profile

Name of the Firm	position of interviewee	Work experience	Interviewee's Education level
Kisasa Supplies Limited	Manager	4 years	Degree
Uncle Milo sunflower Cooking Oil	Accountant	10 years	Higher diploma
Nyemo Investment	Director	10 years	Degree
Ringo Consolidated Company Limited	Director	10 years	Degree
Furaha Dodoma Oil Mills	Manager	1 year	Higher diploma
The 3sisters	Manager/Owner	10 years	Secondary school
Edith Mills	One of owner	10 years	Degree
Pasua Mini-super market	Manager	5 years	Secondary school

2.3.2 Data analysis

This study seeks to investigate how oil processors in Tanzania uptake induced innovation by downstream actors in Tanzania. To analyze qualitative data this study used thematic framework analysis. There are a number of ways of doing thematic framework analysis. The selection of which style to use depends on the basis of the theory and text itself or on the basis of both [57]. Coding framework thematic analysis is based on established criteria. Here the researcher can specify the topic or words. Framework analysis is based on the emerging of the pattern on the determined elements. By using literature on trust, network theory, and innovation, different parameters have been identified as being relevant for the analysis of downstream induced innovation in Tanzania as the guide for the framework. The techniques for analysis of qualitative data can be used in interview script or reports. Attride-Stirling identifies four stages for content framework thematic analysis which are [58]:
1. developing a coding schedule
2. organizing the coded text as themes
3. establishing the common themes
4. analyzing the theme provided by the basic coded material

This paper categorized text into code sentences, basic themes, organizing theme and global theme as shown in Table 2.2.

2.3.3 The analysis step

2.3.3.1 Developing coding materials

Qualitative thematic framework analysis is better adapted to research that has a specific questions, a limited time frame, a predesigned sample and prior issues [59]. This paper used framework qualitative thematic analysis to analyze the uptake of induced innovation by SMEs in Tanzania. The procedure is good for analyses of the comments of the people whom have been affected by the new system of food distribution that is rising in Africa.

Codes: A total of 23 codes as shown in Table 2.2 were developed. Due to the nature of the study some codes appeared in more than one global theme that we developed. Predetermined words used for identification of the codes were:
- *Trust*, including assets, loan repayment records and reputation. Focus is on all activities performed by retailers which affect their reputation among actors.
- *Product innovation*, including products attributes, how they get information and what they have done.
- *Organization innovation*, including sales, loans, distribution policies (regulation).
- *Technological innovation*, including kinds of technology, access to technologies, and support in accessing technologies.

Basic theme: After reduction of the text, the data that have been derived as the codes is assembled together as the basic theme. Attride-Stirling indicates that although it is simply the renaming of the codes it is still crucial and helpful for the creation of the thematic framework [60]. Table 2.2 shows seven basic themes developed from the codes.

Organizing theme: These are themes clustered together from many issues in the basic themes. Table 2.2 shows some of organizing themes that emerged after the joining of the basic themes. After identifying the themes and examining the underlying issues the organizing themes were named.

Global theme: The global theme is the summary of the claim, proposition, argument, assertion or assumption of the organizing theme [61]. The global theme simply reflects ideas behind organizing themes. However, to understand what global themes represent one has to read from global to the basic theme, then coded data. Therefore, for interpretation purposes the author goes back to the data again to understand the meaning of the global theme in a particular context.

2.4 Findings

2.4.1 Organizational innovation

This study finds that SMEs perceive organizational innovation is not an activity that they could do to improve their various activities despite a number of challenges they face. For instance, SMEs in oil processing face non-loan repayment. Despite this, most of them have no guidelines on lending, such as how much and what amount is the limit borrowers can take. Further SMEs perceive business guidelines that would control and improve their performance have to be formulated by central government. One respondent said:

> "You know business policy is so wide. Policy is something to be formulated and implemented by the government."

This has been so because most of the SMEs in food processing believe innovation to be a breakthrough and not an incremental process which would improve their day-to-day activities in their field.

2.4.1.1 Production innovation

Market dynamics have significant influence on changes in production innovation. In general, the study found that government agencies, final consumers and distributors create pressure on sunflower oil processors to adopt or introduce production innovation.

Table 2.2. Thematic Framework network (from codes to global theme)

Code sentences	Basic themes	Organisation theme	Global theme
– We don't trust all information – I don't have to trust anyone – Even financial systems don't trust anyone...why should I trust somebody	Don't trust each other	Lack of trust	Skeptical about market innovation ideas
– We don't trust all information – Telephones – Mail – Social network is for gossip – Mobile phones are used to search for market price	Means of communication Don't trust market information	Market information	Less use of means of communication
– Consumer demand – Higher learning institutions – Retailers opinion	Source of innovation	Product information	Market driven innovation
– Tanzania Bureau of Standards mark – Sealed products – Refinery process – Modern consumers – They are looking for labels	Packaging Branding Food processing	Product attributes	Process innovation
– Law does not constrain us but directs us what to do – No sell policy... policy is very wide and has to be done by the government – No I don't have any guidelines of distribution – Yes some consumers don't pay us...in many times – Last year I got loss of (Tsh*) 3 million – We don't keep records...so we don't understand how much but on average loss amount to (Tsh*)1.5 million – We have all the receipts and every day we are saying will write but we don't	Firm's regulation	Country regulation	Lack of organizational innovation

Note: Tsh stands for Tanzania shilling added by interviewers for clarification

For example, SMEs changed their production systems by separating the packaging unit far from the processing unit after inspection by government agencies involved with food and standards (Tanzania Foods Drugs Authority [TFDA] and Tanzania Bureau of Standards [TBS]). However, some of them introduced aluminum galvanized tanks as storage with pipes that pumped crude processed oil from the processing unit to the packaging unit. This simple innovation, used by all interviewed processors, intends to reduce the contamination of processed oil and limit direct human contact with processed oil.

Empirical evidence shows firms in developing countries can introduce technological innovation by two major means: (i) by learning through the multinational corporations (MNCs) activities, and (ii) by linkage through universities.

Vertical learning through MNCs depends much on spillover effects of foreign direct investment, but this kind of thinking assumes that there is no internal innovation generation in developing countries. Dantas et al. suggest that for any firms that benefit from university linkages innovation absorptive capacity is very important [62].

2.4.1.2 Uptake ability

In general, research on SMEs and innovation agree that the ability of SMEs to introduce innovation is hindered by a lack of funds. However, SMEs can introduce other forms of innovation such as organization and production innovation. But to manage to initiate and introduce other forms of innovations which are not breakthrough innovation ability of uptake is very important.

Innovation uptake ability of SMEs in the value chain from downstream act overs in the value chain has been investigated in this study. In general this study has indicated that SMEs in food processing in Tanzania are driven to introduce innovation, and in particular technology, as a response to government watchdog initiatives.

Absorptive capacity does not only include the ability to introduce new technology, but for SMEs also the ability to fight pirated products. We asked SMEs how they reduce the chances of their products being pirated. One of the interviewees said:

> "I face this problem. Last time one of the distributors in Dar-Es-Salaam complained about the quality of our products. But, we identified it was not my products although the brand was mine. Generally I cannot fight pirated brands of my cooking oil product and I know some people are using that weakness."

2.4.1.3 Lack of trust among actors in the value chain

One of the areas that reduces the uptake of innovation by processors induced by downstream actors in the sunflower processing industry in Tanzania is lack of trust among players in the value chain. Our study indicates that processors don't trust information received from distributors on market information and attributes demanded by final consumers. This is similar to the findings of Gellynck and Kühne, who found that SMEs food manufacturers have no collaboration with retailers, which limits their chances to access information and knowledge to support their innovation initiatives [63]. However, processors learn horizontally and not from actors in the vertical food chain. One of the interviewees said:

> "We bought this machine when one of the processor introduced this machine."

The delay of processors of sunflower oil to introduce new innovation in Tanzania can be attributed to many reasons, such as lack of a stable market, availability of raw materials, inadequate working capital, and reliability of the kind of technology to be adopted. For instance, SMEs interviewed were reluctant to adopt new technology made locally in the sunflower refinery process due to lack of assurance if the new machine complies with TBS standards. One of the interviewees said:

> "I am not sure if TBS would allow this technology invented by VETA. Which can be used for cooking oil refinery? Therefore, I better not use this machine."

2.4.1.4 Reluctance to internally originated new ideas

One of the major sources of innovation for any firm is internal employees. This kind of innovation can be in different forms, such as breakthrough or incremental. However, African employees are afraid to suggest new ideas for the improvement of their performance for fear of losing their jobs [64]. Further, the author argues that the African culture of not telling what you think is good to your seniors. Fear of how they may react limits the exchange of ideas and learning of new ideas in companies in the continent.

This study has found that SMEs are reluctant to adopt new aides induced by internal employees. This can be attributed to two major reasons: (i) fear of making loss, and (ii) lack of adequate knowledge of new ideas. One interviewee said:

> "It's our technicians who insisted we use s pump and install s pipe to allow raw oil to flow from the processing unit to the packaging room. This costs a lot of money but reduces contamination in the produce. Took us some time before we decided to implement an idea which we have today."

2.4.1.5 Influence of external organizations

One of the major institutions that support innovation in Tanzania to be adopted by SMEs is universities. However, there is a contrast in the participation of the higher learning institutions in facilitating the uptake of innovation in Tanzania. De Bruijin and Mahemba found there is weak involvement of universities [65], while Musambala and Azatyan et al. found universities have strong influence on SMEs' innovation uptake as external actors [66, 67]. This study revealed that SMEs' innovation uptake is facilitated by universities, which have a high role in innovation uptake. One of the interviewees said:

> "Major source of innovation includes universities which we are working very closely with."

With the rise of retailers as drivers of value chain in Africa, it's likely that the number of intermediaries have increased in the spread of innovation in developing economies. For example, this study finds that oil processors receive opinions from retailers. But the study also reveals that oil processors are skeptical of the information they receive from actors on the value chain for technological or organization upgrading in Tanzania. Diyamett and Wangwe found that there are few linkages between technological institutions and manufacturing firms at any level in Tanzania [68]. This correlates with our finding that SMEs are skeptical of innovation ideas from downstream or any external actors.

2.5 Conclusion

This study intends to investigate the uptake of innovation induced by downstream actors by sunflowers SMEs in Tanzania in response to an appeal by Beuleans et al. [69]. This study reveals that SMEs pay little attention to organizational innovation. However, they are very focused on production innovation with interest in servicing the market. Further, the study reveals that major factors that influence the uptake of innovation from downstream are not final consumers but government agencies. For instance, most of the SMEs involved in the study search for funds to buy small oil refinery machines after government bans of the selling of raw oil. This suggests that processors still think consumers will absorb everything because there is lack of supply and demand is high.

However, the flood of imported foods in the local market indicates that consumers in Tanzania are no longer absorbing what is found at the market. Instead, they are more selective, which implies that for a firm to survive it has to be more innovative. Nevertheless, SMEs are ambitious to meet the needs of market by inducing different innovation, but this study found that they don't trust information received from downstream players. This limits their speed in implementing various concepts advised by the final consumers through distributors. We presume this was fueled by the shorter kind of relationship between actors in the value chain. Another important finding in this study is that SMEs in Tanzania don't pay attention to organizational innovation that may speed up the distributions of their produce. This finding is similar to that of Mutambi, who finds that SMEs in Uganda sideline organizational innovation and pay attention to production and technological innovation [70].

This study investigated SMEs' uptake of downstream induced innovation in the sunflower value chain. This study used qualitative analysis, and we argue for more qualitative study, such as case studies, to investigate the effect of SMEs accepting induced innovation from the downstream market in other agri-processing sectors. Case studies would allow us to learn the effect of successful and collapsed firms after acceptance of induced downstream innovation. A previous study by Kuada investigates the impact of power and influence on acceptance of innovation in biscuit marking in Ghana [71]. However, this study focused on snacks for export market, whereas we need to understand how firms in agro-processing diversify after unbalanced power relations affect induced innovation, and how they adapt to this shock market change following the failure after the rejection of the downstream buyers. Further study can be done on the capacities of African SMEs to absorb downstream activities.

Further research can also be on the involvement of external actors on the introduction and inducement of innovation in Africa. For example, some innovation is with the initiatives of NGOs in agri-processing, but what influence do emerging retailers have? The current study has tried to identify this, but we need more study on this, particularly on international and local retailing firms operating in Africa. More study has to be done on the effects of externally versus internally initiated innovation on the impact of development of SMEs. This can be on different perspectives from entrepreneurship, management, development economics and finance.

Previous studies in Tanzania have contrary findings on innovation and performance of firms. Kristiansen, Knudsen et al. found there is a link between innovation and firm success [72, 73]. However, De Bruijin and Mahemba found there is a weak link between firm performance and innovation in Tanzania [74]. This suggests that there is more research needed and future research has to be qualitative so that we can learn how entrepreneurs respond to the poor performance despite investigations they have made into innovation in Tanzania. Further, this study has to concern certain commodities instead of having a basket number of SMEs from the food sector [75]. Focus has to be on a specific commodity's value chain for a clear understanding of innovation uptake and diffusion of the actors vertically and horizontally. One of major limitations in this study is lack of information from consumers on the innovation induced by them, if it has been implemented by SMEs, and how they perceive those changes.

Bibliography

[1] Beuleans, A. J. M., Hagen, J. M., Omta, S. W. F. and Trienekens, J. H. Innovation through (international) food supply chain development: a research agenda. Int. Fod. & Agr. Mng. Rev. 6(1) (2003) 86–98.

[2] http://www.mtmerugroup.com.

[3] http://www.corporate-digest.com/index.php/tanzania-spends--120-million-to-import-edible-oil-every-year, (accessed 20.07.2014).

[4] The Guardian: editorial, (Tanzania) Fresh look needed on edible oils production, 28th July 2011.

[5] Corporate digest Tanzania spends $ 120 million to import edible oil every year. http://www.corporate-digest.com/index.php/tanzania-spends--120-million-to-import-edible-oil-every-year, 2013 (accessed 20.07.2014).

[6] Kuada, J. Impact of social ties on innovation and learning in the African context. In: Putting Africa first: The making of African innovation systems, Muchie, Gammeltoft, Lundvall, eds. (Aalborg University Press, 2003, pp. 109–121).

[7] De Bruijin, E. J. and Mahemba, C. M. Innovation activities by small and medium-sized manufacturing entreprises in Tanzania. Cret. & Inn. Mng 12(3) (2003) 162–173.

[8] Musambala, M. Diffusion of innovation in Tanzania rural food processing ventures: a case of sunflower and palm oil, 2012.

[9] Gellynck, X., Kuhne, B., Lefebvre, V. and Vermeire, B. Measuring innovation paucity in the agrifood sector: from single companies to value chains. Jon. on Ch. & Net. Sci. 10(3) (2010) 145–157.

[10] Beuleans, A. J. M., Hagen, J. M., Omta, S. W. F. and Trienekens, J. H. Innovation through (international) food supply chain development: a research agenda. Int. Fod. & Agr. Mng. Rev. 6(1) (2003) 86–98.

[11] Omar, O. E. Retail influence on food technology innovation. Inte. Jon. of Rtl. & Dst. Mng. 21(1) (1995) 11–16.

[12] Grunert, K. G., Jensen, B. B., Sonne, A., Brunsø, K., Byrne, D. V., Clausen, C., Friis, A., Holm, L., Hyldig, G.,Kristensen, N. H., Lettl, C. and Scholderer, J. User-Oriented Innovation in the Food Sector: Relevant Streams of Research and an Agenda for Future Work. Tnd. in Fod. Sci. & Tec. 19 (2008) 590–602.

[13] Beckeman, M., Bourlakis, M. and Olsson, A. The role of manufacturers in food innovations in Sweden. Brit. Fod Jon 115(7) (2013) 935–974.

[14] Schumpeter, J. A. The theory of economic development. (Cambridge, Massachusetts Harvard University Press, 1934).

[15] Schumpeter, J. A. The theory of economic development. (Oxford, Oxford University Press, 1912).

[16] Reardon, T. and Weatherspoon, D. The rise of supermarkets in Africa implications for agrifood systems and the rural poor. Dev. Pol. Rev. 21(3) (2003) 333–355.

[17] Gellynck, X., Kuhne, B., Lefebvre, V. and Vermeire, B. Measuring innovation paucity in the agrifood sector: from single companies to value chains. Jon. on Ch. & Net. Sci. 10(3) (2010) 145–157.

[18] Gellynck, X. and Kühne, B. Innovation and collaboration in traditional food chain networks. Jon on Chn & Net. Sci 8 (2008) 121–129.

[19] Omta, S. W. F. Innovation in chains and networks. Jon. on Ch. & Net. Sci. 10(3) (2002) 73–80.

[20] Agarwal, A. and Shankar, R. Online trust building in e-enabled supply chain. Sup. Chn Mng: an Int. Jon. 8 (2003) 324–334.

[21] Njihoff-Savvaki, R., Omta, S. W. F. and Trienekens, J. H. Drivers for innovation in niche pork netchains: a study of United Kingdom, Greece, and Spain. Brit. Fod Jon 114(8) (2012) 1106–1127.

[22] Ussman, A., Franco, M., Mendes, L. and Almeida, A. Are SMEs Really Innovative? A Study Regarding the Main Difficulties in Portuguese SMEs. Conference Paper No. 78, Conference of the International Council for Small Business (ICSB), Small Business Advancement National Center, Naples, Italy. 1999.

[23] Dabas, C. S., Mahi, H. and Sternquist, B. Organized retailing in India: upstream channel structure and management. Jon. of Bus. Ind. Mkt 27(3) (2012) 176–195.

[24] Scozzi, B., Garavelli, C. and Crowston, K. Methods for modelling and supporting innovation processes in SMEs. Eur. Jon. Inn. Mng. 8 (2005) 120–137.

[25] Twiss, B. The management of technological innovations, 2nd edn. (Programme Press, 1980).

[26] Scozzi, B., Garavelli, C. and Crowston, K. Methods for modelling and supporting innovation processes in SMEs. Eur. Jon. Inn. Mng. 8 (2005) 120–137.

[27] Gellynck, X., Vermeire, B. and Viaene, J. Innovation in food firms: Contribution of regional networks within the international business context. Ent. & Reg. Dev. 19 (2007) 209–226.

[28] Ussman, A., Franco, M., Mendes, L. and Almeida, A. Are SMEs Really Innovative? A Study Regarding the Main Difficulties in Portuguese SMEs. Conference Paper No. 78, Conference of the International Council for Small Business (ICSB), Small Business Advancement National Center, Naples, Italy. 1999.

[29] Pol, H. and Visscher, K. The influence of power in supply chain innovation: a case study of the Dutch wheat chain. Jon on Chn & Net. Sci. 10(1) (2010) 77–85.
[30] Li, H., Wu, X. and Zheng, S. Network resources and the innovation performance: Evidence from China manufacturing firms. Mng. Dec. 51(6) (2013) 1207–1224.
[31] Gellynck, X., Vermeire, B. and Viaene, J. Innovation in food firms: Contribution of regional networks within the international business context. Ent. & Reg. Dev. 19 (2007) 209–226.
[32] Omar, O. E. Retail influence on food technology innovation. Int. Jon. of Rtl. & Dst. Mng 21(1) (1995) 11–16.
[33] Eisenhardt, K. M. and Martin, J. A. Dynamic capabilities what are they?. Str. Mng. Jon. 21 (2000) 1105–1121.
[34] Beckeman, M., Bourlakis, M. and Olsson, A. The role of manufacturers in food innovations in Sweden. Brt. Fod. Jon 115(7) (2013) 935–974.
[35] Ghauri, P. N. and Kemp, R. G. M. Interdependency in joint ventures: the relationship between dependence asymmetry and performance. Jon on Chn & Net. Sci. 1 (2001) 101–110.
[36] Li, H., Wu, X. and Zheng, S. Network resources and the innovation performance: Evidence from China manufacturing firms. Mng. Dec. 51(6) (2013) 1207–1224.
[37] Gellynck, X., Vermeire, B. and Viaene, J. Innovation in food firms: Contribution of regional networks within the international business context. Ent. & Reg. Dev. 19 (2007) 209–226.
[38] Njihoff-Savvaki, R., Omta, S. W. F. and Trienekens, J. H. Drivers for innovation in niche pork netchains: a study of United Kingdom, Greece, and Spain. Brit. Fod Jon 114(8) (2012) 1106–1127.
[39] Omar, O. E. Retail influence on food technology innovation. Int. Jon. of Rtl. & Dst. Mng 21(1) (1995) 11–16.
[40] Kottila, M. Knowledge sharing in organic food supply chains. Jor. on Chn & Net. Sci. 9(2) (2009) 133–144.
[41] Ameseder, C., Fritz, M., Haas, R., Meixner, O. and Schiefer, G. Measurement of the importance of trust elements in agri-food chains: an application of the analytical hierarchy process. Jor. on Chn & Net. Sci. 8(2) (2008) 153–160.
[42] Ghauri, P. N. and Kemp, R. G. M. Interdependency in joint ventures: the relationship between dependence asymmetry and performance. Jor. on Chn & Net. Sci. 1 (2001) 101–110.
[43] Reardon, T. and Weatherspoon, D. The rise of supermarkets in Africa implications for agrifood systems and the rural poor. Dev. Pol. Rev. 21(3) (2003) 333–355.
[44] Kimeme, J., Kristiansen, S., Mbwambo, A. and Wahid, F. Information flows and adaptation in Tanzanian cottage industries. Ent. & Reg. Dev: An Int. Jor.17 (5) (2005) 365–388.
[45] Dabas, C. S., Mahi, H. and Sternquist, B. Organized retailing in India: upstream channel structure and management. Jor. of Bus. Ind. Mar. 27(3) (2012) 176–195.
[46] Dantas, E., Giuliani, E., Marin, A. The persistence of 'capabilities' as a central issues in industrialization strategies: How they relate to MNC spillovers, industrial clusters and knowledge networks. Asi. Jor. of Tec. Inn. 15(2) (2008) 19–43.
[47] RLDC. Sunflower sector: market development strategy, 2008.
[48] Kuada, J. Power asymmetries and relationships between MNCs and the local firms in Africa. Afr. Jor. of Bus. & Eco. Res. 3(2) (2008) 92–105.
[49] Gellynck, X., Kuhne, B., Lefebvre, V. and Vermeire, B. Measuring innovation paucity in the agrifood sector: from single companies to value chains. Jor. on Chn & Net. Sci. 10(3) (2010) 145–157.
[50] Reardon, T. and Weatherspoon, D. The rise of supermarkets in Africa implications for agrifood systems and the rural poor. Dev. Pol. Rev. 21(3) (2003) 333–355.
[51] Kimeme, J., Kristiansen, S., Mbwambo, A. and Wahid, F. Information flows and adaptation in Tanzanian cottage industries. Ent. & Reg. Dev: An Int. Jor. 17 (5) (2005) 365–388.

[52] Ndyetabula, D. W. Growing food business in a development setting: A value chain perspective. (Unpublished PhD Thesis presented at Aalborg University, 2012).
[53] Musambala, M. Diffusion of innovation in Tanzania rural food processing ventures: a case of sunflower and palm oil, 2012.
[54] RLDC. Sunflower sector: market development strategy, 2008.
[55] Kuada, J. Impact of social ties on innovation and learning in the African context. In Muchie, Gammeltoft, Lundvall, eds. Putting Africa first: The making of African innovation systems. (Aalborg University Press, 2003, pp. 109–121.
[56] De Bruijin, E. J. and Mahemba, C. M. Innovation activities by small and medium-sized manufacturing entreprises in Tanzania. Cre. & Inn. Mng. 12(3) (2003) 162–173.
[57] Attride-Stirling, J. Thematic network: an analytic tool for qualitative research. Qua. Res. 1(3) (2001) 385–405.
[58] Attride-Stirling, J. Thematic network: an analytic tool for qualitative research. Qua. Res. 1(3) (2001) 385–405.
[59] Thomson, S. B. and Srivastava, A. Framework analysis: A qualitative methodology for applied policy research. JOAAG 4(2) (2009) 72–78.
[60] Attride-Stirling, J. Thematic network: an analytic tool for qualitative research. Qua. Res. 1(3) (2001) 385–405.
[61] Attride-Stirling, J. Thematic network: an analytic tool for qualitative research. Qua. Res. 1(3) (2001) 385–405.
[62] Dantas, E., Giuliani, E., Marin, A. The persistence of 'capabilities' as a central issues in industrialization strategies: How they relate to MNC spillovers, industrial clusters and knowledge networks. Asi. Jor. of Tec. Inn. 15(2) (2008) 19–43.
[63] Gellynck, X. and Kühne, B. Innovation and collaboration in traditional food chain networks. Jor on Chn & Net. Sci. 8 (2008) 121–129.
[64] Kuada, J. Power asymmetries and relationships between MNCs and the local firms in Africa. Afr. Jorn. of Bus. & Eco. Res. 3(2) (2008) 92–105.
[65] De Bruijin, E. J. and Mahemba, C. M. Innovation activities by small and medium-sizedmanufacturing entreprises in Tanzania. Cret. & Inv. Mng 12(3) (2003) 162–173.
[66] Azatyan, R., Chaminade, C. and Szogs, A. Building absorptive capacity in less developed countries: The case of Tanzania. (CIRCLE, Working paper, no. 2008/05).
[67] Musambala, M. Diffusion of innovation in Tanzania rural food processing ventures: a case of sunflower and palm oil, 2012.
[68] Diyamett, B. and Wangwe, S. Innovation indicators within Sub-Saharan Africa. A specific case for Tanzania. in Blankey, Scerri, Molotja, Saloojee, eds., Measuring Innovation in OECD and Non-OECD countries. (selected seminar paper, HSRC Press, South Africa, 2006).
[69] Beuleans, A. J. M., Hagen, J. M., Omta, S. W. F. and Trienekens, J. H. Innovation through (international) food supply chain development: a research agenda. Int. Fod. & Agr. Mng. Rev. 6(1) (2003) 86–98.
[70] Mutambi, J. Stimulating industrial development in Uganda through open innovation business incubators. (Unpublished PhD thesis submitted at Bleking Institute of Technology, 2011).
[71] Kuada, J. Power asymmetries and relationships between MNCs and the local firms in Africa. Afr. Jorn. of Bus. & Eco. Res. 3(2) (2008) 92–105.
[72] Kristiansen, S. Promoting African pioneers in business: What makes a context conducive to small scale entrepreneurship? Jorn. of Entr. 10(1) (2001) 43–69.
[73] Knudsen, H., Kristiansen, S. and Nsana, B. Apprenticeship and entrepreneurial development in the Tanzania informal sector. (Adger Collge/IDM collaboration research report no. 6, Mzumbe University Tanzania, 1994).

[74] De Bruijin, E. J. and Mahemba, C. M. Innovation activities by small and medium-sizedmanufacturing entreprises in Tanzania. Cret. & Inv. Mng 12(3) (2003) 162–173.
[75] De Bruijin, E. J. and Mahemba, C. M. Innovation activities by small and medium-sizedmanufacturing entreprises in Tanzania. Cret. & Inv. Mng 12(3) (2003) 162–173.

Aykut Berber

3 Customer Experience, Technology and Innovation: Evidence from Georgian London and the Victorian Era

Abstract: Although mainstream innovation literature emerged on economic and industrial grounds, it began to expand dramatically in the Information Age. However, the concept of innovation is as old as the humankind by its nature; and history reveals many successful innovations realized by talented entrepreneurs and pioneers of their respective times. In this context, this chapter depicts two cases in different settings. The first case illustrates a picture of the coffeehouses in Georgian London and examines how an unfamiliar commodity like coffee built strong ties between people and inspired coffeehouse owners like Edward Lloyd to generate more experience-driven innovations. The second case takes place during the Victorian era and expands on the entrepreneurial mindset of William Davidson, the pioneer of the refrigerated food shipping industry, who decisively made use of an unfamiliar technology and managed to export frozen meat from New Zealand to London.

3.1 Georgian Londoners and Victorian Entrepreneurs

After opening the *Great Exhibition of the Works of Industry of all Nations* on 1 May 1851, Queen Victoria wrote these words in her diary: "This day is one of the greatest and most glorious of our lives" [1]. And it literally was. The Queen and all other visitors were fascinated by the Crystal Palace which was built on the grounds of Hyde Park especially for this international exposition. The Great Exhibition remained open for nearly five months and received more than six million visitors. One of them was Charlotte Brontë, the famous author, who described the venue and the exhibits as follows:

> It is a wonderful place – vast, strange, new and impossible to describe. (...) Whatever human industry has created you find there, from the great compartments filled with railway engines and boilers (...) to the glass-covered and velvet-spread stands loaded with the most gorgeous work of the goldsmith and silversmith, and the carefully guarded caskets full of real diamonds and pearls worth hundreds of thousands of pounds (p. 216 in [2]).

The Exhibition itself was great proof of the reality that arts, science and industry were not separable. Exhibits from Britain and the Colonies occupied half of the venue leaving the other half to exhibitors that came to London from the rest of the world. Among

Aykut Berber: Istanbul University School of Business, Istanbul, Turkey and visiting professor (Spring 2014) at Brunel University Brunel Business School, London, UK. berber@istanbul.edu.tr

the exhibits were steam-powered machinery, a giant model of Liverpool docks, and the electric telegraph (p. 69 in [3]) along with gold watches from Switzerland; porcelain, silks, tapestries and furniture from France; McCormick's reaping machine, Goodyear rubber goods and Colt's fire-arms from the United States; the Koh-i-Noor diamond (which was the largest in the world at the time) from India; and many more from other countries [4]. Charles Babbage, mathematician and father of the programmable computer, was among the Exhibition's closest witnesses. Being concerned with the decline in the development of labor-saving machinery in Britain and predicting a lack in adopting science into technology (pp. 223–224 in [5]), Babbage published his thoughts and critiques in *The Exhibition of 1851* the same year as the Exhibition was opened. This is how he described the purpose of the Exhibition in the Preface of the book:

> England has invited the civilized world to meet in its great commercial center; asking it, in friendly rivalry, to display for the common advantage of all, those objects which each country derives from the gifts of nature, and on which it confers additional utility by processes of industrial art. This invitation, universally accepted, will bring from every quarter a multitude of people greater than has yet assembled in any western city: these welcome visitors will enjoy more time and opportunity for observation than has ever been afforded on any previous occasion. The statesman and the philosopher, the manufacturer and the merchant, and all enlightened observers of human nature, may avail themselves of the opportunity afforded by their visit... (Preface in [6]).

1851 may be assumed as the peak year of an era where the British enjoyed the fruits of the global trade and the Industrial Revolution. In fact, the Georgian and Victorian eras in British history reveal much about the history of a society which witnessed and experienced remarkable changes as a collective consequence of a great sociopolitical paradigm shift, numerous technological developments, and economic growth. It spanned two centuries that saw great events and incidents that were respectively radical and quite exceptional. The United Kingdom and the United States of America came into being; Adam Smith published *The Wealth of Nations*; Richard Arkwright established arguably the world's first factory in Derbyshire; Josiah Wedgwood successfully industrialized the pottery artwork; George Frideric Handel introduced Italian operas and oratorios to the English audience; the East India Company became a huge power in international trade; Stamford Raffles founded Singapore; the steam engine was invented; canals and bridges were built; railways were constructed; slavery was abolished – just to name a few examples. The thriving economy transformed Britain into the world power of the time and the British society enjoyed investing in new ventures and experiencing various tastes and traditions from around the world.

The diversity of commodities and products in Georgian London was an outcome of the mercantilist policies that controlled the flow of goods and protected the rights of the merchants. The systematic and well-organized trade contributed much to the establishment of regular businesses and to the consumption of goods available at all times. Undoubtedly, fulfilling fundamental needs was still the main purpose of work; however, working profitably and spending wisely also became an individual ability that would eventually produce a society of not mere buyers but customers who attributed value to the products and enjoyed their consumption. This led to a rise in the number of inns, taverns, as well as coffeehouses and tearooms scattered across the city.

In what sense the Victorian era, that spans the years from 1837 to 1901, can be associated with the Georgian era is a separate debate which is out of our scope, but it is definitely an era of diverse events and remarkable developments. Although there are plenty of studies on the inventions and innovations during the second half of the nineteenth century, it is likely that Victorian entrepreneurship did not receive the deserved attention of business management scholars. However, the noteworthy work of Casson and Buckley reveals essential and invincible key points in this matter. According to the authors, "the evidence of Victorian entrepreneurship is compatible with general theories of entrepreneurship which emphasize the role of entrepreneurs in making sound decisions regarding risky innovations". Unlike the relatively small factory or shop ownerships in Georgian times, Victorian entrepreneurship was based rather on successful partnerships between "wealthy investors and professional specialists" (p. 291 in [7]).

3.1.1 The Question of Newness

In almost every study in the field of innovation research, an emphasis is put on the perception of the user of the product. Innovation, in effect, can be a good or a service, or even an idea that is perceived *new* by its users (p. 81 in [8] & [9, 10]). At this point, another keyword comes into existence that frames our concept in the discussion: *newness*. The newness perceived is not necessarily related to the item's time of creation; but, as emphasized also by Miles (p. 81 in [8]), the newness perceived refers to "the newness of the application for helping address a need or for solving some sort of problem". It is the mind of the people that labels an item as an innovation, and the mind of the innovator needs to predict the correct combination of time, end-users and place to generate this label. As a matter of fact, newness of a product (or a service offered) is a complicated issue. As suggested by Goldenberg and Mazursky, newness to the market is usually assumed to be an advantage for the product success whereas newness to the firm as well as technological change may not provide the same effect (p. 213 in [11]).

Without dwelling much on theoretical issues, we will now attempt to explore how innovation was driven as a consequence of efforts and observations of business owners or managers in the past who seek the best response to the question of newness on two different occasions: introduction of a new commodity to a market, and integration of a new technology into transportation of goods. First, we will go back to Georgian London to explore how an unknown commodity turned out to be so popular as a connection-maker between people rather than a fashionable hot drink to enjoy. Next, we will take a challenging trip from New Zealand to Victorian London to discover how the entrepreneurial mindset of a manager turned out to be so successful in creating a new industry and redefining the rules of a game.

3.2 London Coffeehouses and the Georgians

English chaplain William Biddulph who visited Istanbul in the beginning of the seventeenth century once wrote that coffee was the common drink of Turks, and described this precious commodity as:

> (...) a blacke kind of drinke made of a kind of Pulse like Pease (...) which being grownd in the mill, and boiled in water, they drinke it as hot as they can suffer it [12].

According to Biddulph, coffee suited the Turkish constitution very well, while it did not suit that of the Englishman. As Ellis denotes, "coffee was the sign of Turkish difference" [12], and unsurprisingly Biddulph, as a preacher who accompanied English merchants at that time, could have been rather reluctant to build a cultural connection between the idea of drinking coffee and the existing English customs. However, only a few decades later, England was to discover the taste and joy of this mystic hot drink. Coffee trade began in Oxford in 1650, and arguably the first coffeehouse in London was opened in 1652 [12, 13]. To the English, coffee was new and coffeehouses were neoteric. Compared to alehouses, these little shops acted as cozy places of sociability where people gathered to exchange news and to meet with other people. Although coffee has never been a rival against beer and wine, coffeehouses – on account of the lack of alcohol – also served as perfect centers for engaging in business networks and dealing with serious money issues that necessarily required sobriety. By the end of the century, the number of coffeehouses began to rise. The Queen Anne period and the Georgian era saw the opening of numerous coffeehouses. Even a particular style of coffee table – a light and comfortable piece of furniture – was named after Queen Anne and is still manufactured and sold in stores today. However, it was actually with the beginning of the Georgian era that London became the city of pleasure where many inns, taverns, gardens and of course, coffeehouses filled out the streets (pp. 169–170 in [14]).

Coffeehouses were exotic yet functional places where intellectual debates over philosophy and politics were often seen, social gatherings around particular interests occurred, and even new money-making trends emerged. For instance, Jonathan's coffeehouse in Exchange Alley was the popular place where the South Sea Company shares were intensively traded. The company was established in 1711 by the Earl of Oxford to improve the trade with Spanish America. Relying on the company's anticipated revenues from the trade with the New World, many people were zealous to buy its shares. However, the company went bankrupt and failed to repay its debts (p. 168

in [15]). Isaac Newton and John Gay (who later wrote *The Beggar's Opera*) were among these shareholders (p. 54 in [16]). Alternatively, marine insurers, who used to meet in the Royal Exchange until the Great Fire of 1666, began to crowd the recently opened Lloyd's coffeehouse to exchange and to discuss the latest information on maritime business matters. A few years later, landlord Edward Lloyd (c. 1648–1713) decided to disseminate the news via a special gazette, which still publishes issues under the name of *Lloyd's List* – holding in some sense the title of the oldest running journal in the world (p. 147 in [14]). Conversely, more amusing and exciting events happened in Button's – the favorite coffeehouse of poets and playwrights. Here, a roaring lion's head made of white marble was placed on the wall. By dropping envelopes into the lion's mouth, amateurs could submit their essays and stories with the hope of getting them selected for publication [17] and (pp. 314–316 in [18]).

Many other groups of people other than exchange dealers, stockjobbers, insurers and playwrights crowded the coffeehouses of Georgian London. Booksellers, lawyers, clergymen, artists, authors, military men, and members and supporters of political parties were frequenters of their respective coffeehouses scattered around the capital (p. 170 in [14]). According to the Swiss traveler César de Saussure, who witnessed the diverting growth of London in the 1730s, coffeehouses were not well-groomed and nicely decorated places; besides, they were full of people and full of smoke. Yet, he admitted that the Englishmen were 'newsmongers' as every morning they were anxiously reading the latest news in these shops. As a matter of fact, thanks to the coffeehouses, Georgian London became an essential stage for the emergence of many newspapers that enhanced civil liberty (pp. 170–171 in [14]).

Coffeehouses also played a functional role in the introduction of two more hot drinks to the British market. Only five years after the first coffeehouse was opened in London, "an excellent West India drink" called chocolate was introduced by a French tradesman. In France and Spain, chocolate drink was associated with the social elite, while in London it became available in many coffeehouses [19]. The other hot drink is of course tea which was introduced during the same era and later became eminently associated with the Victorian lifestyle.

Many coffee chain stores as well as boutique coffee shops are scattered across cities of the world today. These entities are usually associated with such concepts as *conversation*, *creativity* and *business*. This shows that coffee still acts as a tie-builder between people sharing a common interest or interacting each other for the exchange of news. Thanks to milk and many other additional ingredients, contemporary shops today introduce many serving options for coffee. However, as coffee was just 'boiled coffee' in Georgian London, it is not surprising that most coffeehouse frequenters may have found it tasteless, and even compared its taste to 'ink', 'soot' and 'mud'! On the other hand, this tasteless yet addictive substance was believed to revive the body and to open the mind [17]. Coffee was a mere product but a coffeehouse was a *platform* for an individual experience.

One can learn a lot from the epic of Starbucks. Without repeating the noteworthy and well-known story here, we can still remember a couple of fundamental ideas that transformed a modest coffee shop in Seattle into a global network of coffee stores. Today, Starbucks and many other coffee stores might be referred to as places of inspiration and innovation where ideas sparkle, business thinkers and designers discuss and dealers negotiate. In fact, at the time of its foundation Starbucks itself was an innovation. It was designed as a creative and profitable platform where frequenters (or guests) would feel as comfortable and relaxed as at home and as motivated and organized as at work. As Howard Schultz, the key figure behind the story, mentions in his book, Starbucks was established due to the need of a 'Third Place' – a place outside of work and home. Referring to German beer gardens, English pubs and French cafés as providers of "a neutral ground where all are equal and conversation is the main activity" (p. 120 in [20]), Schultz continues as follows:

> Most (Americans) just grab their coffee and depart. Still, Americans are so hungry for a community that some of our customers began gathering in our stores, making appointments with friends, holding meetings, striking up conversations with other regulars. Once we understood the powerful need for a Third Place, we were able to respond by building larger stores, with more seating (pp. 121 in [20]).

An experience-driven innovation is a challenging but fruitful process. Customers that benefit from a company's products and/or services are not mere buyers and users but also great potential sources of information for a company that intends to survive in a market. Such information is usually tacit; neither company managers nor customers may directly articulate it. Right on this account, it would not be wise for a manager or a business owner to deny the importance of each individual interaction between the company and its customers. Every individual interaction reveals how the company operates its platform for the individual customer and every individual customer experience possesses essential information. The need of Americans for a community was observed by Schultz, just as the need for the dissemination of news via a special gazette was realized by Edward Lloyd. No matter whether a coffee addict or not, it was and it still is the platform that brought and continues to bring a frequenter to an English coffeehouse or a contemporary coffee store. For each visit, the frequenter anticipates experiencing the ambiance emerged on the platform. Coffee is the central component of this ambiance; in fact it is the nucleus that brings together all other components such as hearing the latest news, discussing political matters, exchanging shares, talking business and chatting with friends.

3.3 Shipping Frozen Meat from New Zealand to London

The mid-nineteenth century saw the scarcity of staple commodities in Britain, and meat was perhaps the most essential of them. The urban population in industrialized cities was growing so fast that food supply was becoming a major problem. On the contrary, the southern hemisphere – particularly Australia and New Zealand – was facing a surplus flocks of sheep which were intentionally grown for their wool. Exporting livestock to the UK was not a practical solution to the surplus problem as the long-distance voyage that took months was quite a challenge to keeping the animals alive on board [21]. In 1869 canning technology was initiated to preserve and export the meat produced in New Zealand. However, it was apparent that canning could not be a solution as only the best part of the carcass could be preserved and the rest was waste. This and many other inefficient attempts to respond to the shortage in Britain continued until refrigeration technology was enhanced enough to transport frozen meat from New Zealand to Britain.

William Soltau Davidson (1846–1924), a Canadian-born entrepreneur from Scotland who was responsible for sheep farming in a land company highly involved in the wool business, was concerned with the inefficient use of the huge flock of sheep of his company. He later became the director of the New Zealand and Australia Land Company (NZALC). As the international wool trade was becoming more threatened and less profitable as a consequence of the hastily multiplying sheep population in Australia and New Zealand, Davidson felt more determined to overcome the difficulties in exporting his company's high quality meat to Britain. An iron ship called Dunedin originally designed for fast passages in the emigrant trade with a capacity of 400 passengers was installed with Bell–Coleman cold-air refrigeration machines. The voyage began in February 1882 with a cargo of around 5,000 frozen mutton and lamb carcasses, and it took Dunedin 98 days to reach Britain's shores. Except for a few incidents on board, the first ever trip for long-distance shipping of frozen meat was successful. The carcasses were immediately brought to London and sold out within two weeks yielding a huge profit ([21–24] and p. 65 in [25]). Davidson's undaunted and determined efforts proved that even companies in countries as remote as New Zealand might take their place in international markets with their products. His creative perseverance made NZALC the pioneer of the frozen meat shipping industry which would soon contribute much to the development of the New Zealand economy while creating a brand new industry in Britain.

The story of the frozen meat shipping industry can arguably be regarded as one of the many proofs of Schumpeter's seminal and prominent arguments on innovation and entrepreneurial activities. The entrepreneur as a change agent is considered to be the person "who makes the untried into a fact"; and "entrepreneurship is innovation and the actualization of innovation" (p. 65 in [26]). Apart from scientific activities and inventions, Schumpeter puts a distinctive emphasis on what we should understand from the concept of innovation as – in his words – "innovation is possible without anything we should identify as invention and invention does not necessarily induce innovation" (p. 84 in [27]). As clarified by Reisman, Schumpeter was concerned with neither scientific novelties nor developments but with "how these developments like electrical power and the motor car become commercialized into a new production function" (p. 65 in [26]). In our case, it is quite evident that what an ordinary consumer in London would have seen as innovation was not how the meat was brought refrigerated and fresh after a three-month voyage from New Zealand, but the outcome of the whole transportation process: "The high quality meat from far away lands at a reasonable price sold right here in the market."

The Dunedin story is a typical Victorian case that reflects the undaunted and persistent characteristic of the post-Industrial Revolution entrepreneurial mindset. The story itself can be investigated like the two sides of the same coin. While the heads side illustrated a picture of the opportunity to create a new industry based on long-distance shipping of frozen meat, the tails side showed the potential challenges caused by the newness of the technology. Even though Davidson and his partners were lucky enough to benefit from the recently developed refrigeration technology, they had to face the ambiguity that emerged because of the many unique problems that were usually inexplicable and compelling. The Bell–Coleman cold-air refrigerating machine that was installed in Dunedin required a steam engine for power. Before the ship's first attempt at departure by the end of 1881, the engine's crankshaft broke and halted freezing. As a result, 641 carcasses that were already loaded and frozen were unloaded and sold in the local market. The departure was possible only after a new crankshaft was made; so the voyage of Dunedin began in February 1882 with new carcasses loaded and frozen on board. Davidson took an earlier trip to London in order to arrange the paper work and organize the distribution of the cargo [28]. The voyage itself was unsurprisingly more ambiguous. The shipping may be assumed to be a modest act of commerce; however the risk that Davidson undertook in solving his company's livestock surplus problem was noteworthy. Here is one of the incidents that proves the risk he had undertaken from Critchell and Raymond's *History of the Frozen Meat Trade* which was published in 1912:

> Captain Whitson ... came on to London ahead of his ship in a pilot boat, looking very strained and careworn as he entered the shipping company's office. He was not quite sure about the condition of the cargo, but thought that most of it was sound. (...) In the tropics the ship was for a long time on one tack, and owing to its steadiness the cold air was not sufficiently diffused amongst the carcasses, and, in fact, the temperature in the upper chamber remained so high that the engineer was almost in despair. At last Captain Whitson had determined to alter the circulation of the air, which was evidently defective, and to do this he had to crawl down the main trunk, and in the process of cutting fresh openings for the better escape of the cold air he became so benumbed by the frost that he was only rescued ... by the mate crawling in behind him and attaching a rope to his legs by which means he was pulled out of the air trunk! (p. 41 in [29]).

On the heads side of the coin is 'the opportunity to reinvent an industry' – as there was already an increasing demand for staple food and particularly for meat in London, the target market. So Davidson's concern was not about introducing a brand new product but it was a question of distribution.

Dunedin's success led to a growth of interest in using steamships for shipping frozen meat from New Zealand [30]. In a sense, what the whole effort caused may be regarded as a disruptive innovation – thanks to a new technology integrated into the distribution channel, a new market was created ([31] and see also: [32–34]). At first, London butchers and salesmen were doubtful about the quality of the frozen meat from New Zealand. However, once they saw that the carcasses were as clean and bright as the fresh British meat, they were convinced enough to find it suitable for the English market and to describe its quality "as perfect as meat could be". The shipment was discussed even in the House of Lords (p. 42 in [29]). It was a turning point in the British meat market that saw the emergence of the frozen meat industry along which would come new ways of brand engagement, new business models, new packaging systems, new manufacturing processes and many other innovations.

3.4 A Final Remark

> Also innate to creativity is inspiration, which has historically been referred to as the *muse* whose influence is often primarily subliminal. Classically, the muse has been associated with the arts, but it is equally a critical factor in the sciences as represented, for instance, by Albert Einstein or the development of quantum physics (Foreword in [35]).

The above text borrowed from Hawkins' foreword to *Creativity Revealed* by Jeffrey reveals, perhaps, the most crucial ingredient of an innovation process: *inspiration*. In this chapter, we might not have expressed much interest in this concept; however it would be unrealistic to deny that all those coffeehouse owners in Georgian London or Davidson in New Zealand were inspired by something. Inspiration has been associated for many centuries with arts rather than technology; in fact it is an intrinsic experience – it is neither inoculated nor installed into one's mind. An unfamiliar commodity like coffee certainly provided Edward Lloyd with a great opportunity to open a coffeehouse yet it must also have caused a big change in his world view. After having observed for a while the people who were frequently coming together to exchange the latest news in his coffeehouse he was inspired enough to start publishing a journal – something perhaps he would never think to do. As for William Davidson, it was all about his environment and his awareness. Apparently, he was inspired to see the opportunity of doing the right business – selling meat – once he realized that he was in the right place at the right time... just as Lloyd was. And just as accurate as a reply to the question Peter Drucker once posed can be: "How much of innovation is inspiration, and how much is hard work?" [36].

Bibliography

[1] The Great Exhibition: Queen Victoria's journal. www.vam.ac.uk. Accessed on 3 July 2014.

[2] Shorter CK. The Brontës; life and letters: Being an attempt to present a full and final record of the lives of the three sisters, Charlotte, Emily and Anne Brontë. Vol. 2. Cambridge University Press, 2013.

[3] Smith N. Events and outcomes; The Industrial Revolution. Evans Brothers, 2009.

[4] Official descriptive and illustrated catalogue of the Great Exhibition. London Spicer Brothers, 1851.

[5] Hyman A. Charles Babbage: Pioneer of the computer. Princeton University Press, 1985.

[6] Babbage C. The Exposition of 1851, or, Views of the Industry, the Science, and the Government of England. J. Murray, 1851.

[7] Casson M, Buckley PJ. Entrepreneurship. Edward Elgar Publishing, 2010.

[8] Miles JA. Management and organization theory: A Jossey-Bass reader. John Wiley & Sons, 2012.

[9] Rogers EM. Diffusion of innovations. Simon and Schuster, 2010.

[10] Rogers EM. A prospective and retrospective look at the diffusion model. Journal of Health Communication 9.S1 (2004): 13–19.

[11] Goldenberg J, Mazursky D. Creativity in product innovation. Cambridge University Press, 2002.

[12] Ellis M. The coffee-house. Hachette UK, 2011.

[13] Suter K. The Rise and Fall of English Coffee Houses. Contemporary Review 286.1669 (2005): 107–110.

[14] Porter R. London: a social history. Harvard University Press, 1998.

[15] Weinberg BA, Bealer BK. The world of caffeine: the science and culture of the world's most popular drug. Psychology Press, 2001.

[16] Robinson EL. Gulliver as slave trader: Racism reviled by Jonathan Swift. McFarland, 2006.

[17] Green M. London cafes: the surprising history of London's lost coffeehouses. http://www.telegraph.co.uk. Accessed on 10 July 2014.

[18] Wheatley HB, Cunningham P. London, past and present: its history, associations, and traditions. Vol. 3. John Murray, 1891.

[19] Green M. The surprising history of London's lost chocolate houses. http://www.telegraph.co.uk. Accessed on 10 July 2014.

[20] Schultz H, Jones Yang D. Pour your heart into it: How Starbucks built a company one cup at a time. Hyperion, 1997.

[21] Tennent K. Management and the free-standing company: The New Zealand and Australia Land Company c. 1866–1900. The Journal of Imperial and Commonwealth History 41.1 (2013): 81–97.

[22] Then and now. nzfarmers.co.uk. Accessed on 8 July 2014.

[23] The nineteenth century heritage: Refrigeration and the meat industry. techhistory.co.nz. Retrieved on 8 July 2014.

[24] Oddy DJ. The Growth of Britain's Refrigerated Meat Trade, 1880–1939. The Mariner's Mirror 93.3 (2007): 269–280.

[25] Ville SP. The rural entrepreneurs: A history of the stock and station agent industry in Australia and New Zealand. Cambridge University Press, 2000.

[26] Reisman DA. Schumpeter's market: enterprise and evolution. Edward Elgar Publishing, 2004.

[27] Schumpeter JA. Business cycles. Vol. 1. New York: McGraw-Hill, 1939.

[28] Some clarifications but yet more questions regarding the early days of the New Zealand frozen-meat trade. The Mariner's Mirror 99.2 (2013): 212–219.

[29] Critchell JT, Raymond J. A history of the frozen meat trade: an account of the development and present day methods of preparation, transport, and marketing of frozen and chilled meats. Constable, 1912.

[30] Murphy H, Oddy D. The business interests of Sir James Caird of Glenfarquhar, Bt (1864–1954). The Mariner's Mirror 97.1 (2011): 22–36.

[31] Yu D, Hang CC. A reflective review of disruptive innovation theory. International Journal of Management Reviews 12.4 (2010): 435–452.

[32] Markides C. Disruptive innovation: In need of better theory. Journal of Product Innovation Management 23.1 (2006): 19–25.

[33] Christensen C. The innovator's dilemma: when new technologies cause great firms to fail. Harvard Business Review Press, 2013.

[34] Christensen CM, Raynor ME. The innovator's solution. Harvard Business Press, 2003.

[35] Hawkins DR. Foreword. In: Jeffrey S. Creativity revealed: Discovering the source of inspiration. Creative Crayon Publishers, 2008.

[36] Drucker PF. The discipline of innovation. Harvard Business Review 63.3 (1985): 67–72.

Manuel Laranja

4 Industrial Resilience: Reframing the Role of Innovation Policies for Regional Development

Abstract: This chapter reviews the literature on industrial and regional innovation policies.

Following on from the debate between those who see the development of Global Value Chains as independent from geographical distance and those that suggest that distance matters and therefore you cannot just access any kind of capabilities anywhere in the world, the paper explores the impacts of these global changes in the international business environment and on regional and multilevel policies for innovation. Our concern is with local responses to globalization through strengthening 'local capabilities for innovation'. We believe that current policy reactions are partial and often distorted by a misconception of the role of local learning and innovation in international regional competitiveness. Therefore we propose to rethink regional and multilevel innovation policies in order to better reflect this change and provide governments with policy tools that effectively help regions to create and protect their capabilities while at the same time being connected to global markets.

4.1 Introduction

In the last few decades a combination of factors, such as the rapid diffusion of reliable high-speed communication networks, reduced trade barriers, freer flows of capital and reduced transportation costs, amongst others, have enlarged the geographic scope of economic activity and market competition, deeply changing the outlook of the world economy [1].

While a growing number of developing lower-wage countries became significant players in international trade and investment, stronger growth rates in these countries is also creating vast new consumer markets and therefore opening new possibilities for investment in the opposite direction, i.e. from higher-wage to lower-wage economies.

In addition, the wider use of low cost ICT and web technologies, in both developing and more developed countries, has increased the possibilities for dismantling large vertically-integrated operations into networks of business units located in different countries – forming the so called global value chains (GVCs) – each supplying and producing components, to be combined by yet another business unit, into final products.

Manuel Laranja: School of Economics and management, University of Lisbon. Email: mlaranja@iseg.ulisboa.pt

Outsourcing and offshoring of activities has been growing, especially in manufacturing industries where product and production processes' modularity is high [2], but also in service sectors such as retailers and branded merchandisers with little or no internal production [3–5].

The development of these so-called GVCs is, however, often associated with the decline of employment in manufacturing sectors located in the developed world. In fact, following a trend to diminish, the weight of manufacturing in most developed OECD countries represents today less than 20% of GDP.

But while these trade and investment linkages between developed and developing countries are not new, their scale and complexity has substantially increased, triggering much debate on the characteristics of this new environment for international economics and on the role of regional innovation policies in promoting local employment, growth and competitiveness.

On the one hand we have those who see the world being 'flat' [6] – or the world without borders [7]. The argument of a flat world means that distance does not matter. You can access production and services capacity anywhere in the world. Also, it is not just production capacity at relatively lower wage costs that you can access worldwide, but also research and development (R&D) and technology. Increasingly we find R&D delocalization and growing investments in global R&D and innovation [8].

In line with international product life-cycle theories [9] and standard trade theories of comparative advantage, this line of argument sees the decline of manufacturing as a natural and healthy evolution. The sociologist Daniel Bell in 1973 [10], has even proposed that the decline in manufacturing in the more advanced economies is a natural and healthy transition to a 'post-industrial' society dominated by 'knowledge workers'.

Delocalization of production to lower wage regions is taken as a healthy symptom of economic development, 'liberating' resources so that they can be used in high-value-added sectors, such as services. Moreover, local production is dispensable if you want to move forward towards a knowledge-based economy focused on services and innovation. To enter so-called 'post-industrial society' or a globally networked society supported by global high-speed digital networks [11], regions should exclude direct manufacturing and rely exclusively on being a hub for high-added-value activities such as R&D, design, testing, or perhaps concentrate on the development of mobile and web applications, or even more recently to focus on promoting the so-called 'creative industries'.

On the other hand, we have those that suggest that distance matters and therefore you cannot just access any kind of static or dynamic capabilities anywhere in the world. The argument is that distance matters for the creation of localized learning linkages between R&D, design, testing and manufacturing. Proponents of this view argue that building infrastructure, improving educational performance, and strengthening cooperation between public and private institutions is better undertaken at the local level.

Following on from this debate, this paper explores the impacts of these global trends on regional innovation systems and on regional policies for innovation. Our concern is with local responses to globalization through strengthening 'local capabilities for innovation'. By 'capabilities for innovation', we mean the ability to conceive, develop, and/or produce and commercialize new products and services. In particular we focus on the contribution that local learning linkages between manufacturing and industrial design and product development make to the construction of regional capabilities, serving the regional innovation ecosystem.

In Section 4.2 we explore in more detail the characteristics of distributed global value chains. Then, in Section 4.3, we will look at the literature on the influence of the close environment in business innovation and competitiveness. This will lead us to the need to reframe the industrializations debate in Section 4.4. Finally in Section 4.5 we attempt to draw some conclusions for a regional and multilevel innovation policy agenda.

4.2 A world of 'distributed' capabilities in global value chains

As hinted above deverticalization of 'manufacturers' that have shed internal capacity and have come to rely on an emergent set of global and East Asian contract manufacturers for production appears to be a natural trend of developed economies [5, 12].

'Traditional' vertically integrated companies may today choose to be 'networked companies', i.e. they may choose to focus and specialize on certain stages or functions of the GVC, while outsourcing extensively to suppliers and subcontractors located in the region which does it best at the lower cost.

The well-known case of the global smartphones industry is illustrative of how in many industries the GVC became truly global. However, although many of these subcontractors are now able to supply complex parts and subsystems to anywhere in world, so far they capture a relatively small portion of the value chain. For example, outsourcing the assembly of the iPhone 4 represents less than 2% of total value, and purchase of materials and components (out of the US) represent around 30% [13].

Stan Shih, chairman of the Taiwan-based Acer Inc, is perhaps the first to have coined the term 'smiling curve' – a U-shaped value distribution across the value chain. In the global smartphones industry, as in personal computers and many other industries, value added is higher in R&D, design and branding, and lower in the midstream stages, which involve labor-intensive processes such as assembly. Because such processes capture only a minor proportion of value added, it is assumed they can and should be outsourced to lower wage regions.

We used to think about local manufacturing industries as important because of the so-called multiplier effects. Bonvillian [14] estimated that each manufacturing job creates between 2.5 and 2.9 jobs in other sectors and that manufacturing also operates as an output multiplier. Also, not just with mass standardized products, but with adequate use of new technologies and industrial management strategies, other modes of production, namely mass customization or flexible manufacturing, can also be scalable. Finally, manufacturing contributes directly to productivity gains and because of this it is very unlikely that it would sustain large-scale job growth as in the past.

Although the smiling curve clearly overlooks multiplier effects and local linkages, and highlight the 'comparative advantage' of delocalization, reality may however be more complex and difficult to interpret.

Suzanne Berger [15] and her team at MIT, studied the actual experiences of 500 companies in North America, Asia, and Europe as they responded to globalization. The study included firms in slow-tech industries like textile and apparel, where technologies and processes evolve incrementally, and fast-tech industries like electronics, where radical disruptive innovations may often cause profound industrial changes. At least two important conclusions emerged from this study.

First, Berger argues that there is no single best model, i.e. there are many different ways to win in the global economy and opportunities are much wider than usually imagined. In some cases successful companies that were once vertically integrated choose to focus on some specialization, while forming different kinds of worldwide partnerships and networks. In other cases, while remaining vertically integrated they find that selective delocalization brings some kind of competitive advantage.

For example while Samsung, a vertically integrated electronics company, chooses to make almost everything in-house, Dell, an American computer company, focuses its own organization on distribution and outsources all the manufacturing and assembly overseas. In the textile and apparel industries, Gap and Liz Claiborne choose to outsource production to foreign countries, but the fastest-growing retailer is the Inditex Group, a Spanish company that makes more than half its clothing in the Galicia region.

Apparently, the technological and organizational changes of the past decades that made modular production possible across a wide range of industries, combined with rising production capabilities in lower wage developing countries contributed to a multiplication of successful models. It is not the simple geographic spread of economic activities across national boundaries, but the functional integration of internationally dispersed activities that is the key to global competitiveness in these different models [16].

The picture captured by Berger [15] is quite different from that of a flat world where any and every industry can access production, design or development capabilities anywhere. Instead, this new environment created the opportunity for vertically integrated companies relatively concentrated in one region to coexist with network-integrated companies or groups of companies spread worldwide [17].

Second, Berger [15] argues that the strength of the international companies examined is 'grounded' in their local environment. For US companies the key aspects appear to be the establishment of local connections between R&D and end users, strong linkages to venture capital and flexible labor markets. A few years later, in the MIT PIE research study, Berger ([18], p.14) reinforced the importance of 'regionally based resources' such as training, collaborations between firms and universities, suppliers, industrial and technical research centers, and so on. The density, diversity, and abundance of such local resources are key characteristics of successful regional innovation systems.

Another important aspect of the increasing fragmentation of global value chains is that delocalization does not only affect manufacturing. Increasingly it also affects R&D, design, initial tests for proof of concept, and so on. For example, international flows of qualified scientists and researchers and cross-border co-operation in science, technology and innovation, are on the rise, as illustrated by indicators such as expenditures with R&D abroad and co-authorship of scientific publications and patents. There is also higher competition between regions that want to attract the major international scientific centers and often, particularly in smaller regions, the presence of a multinational affiliate may account for a high proportion of local R&D activities.

To summarize, globalization, the rapid spread of new technologies and rising capabilities in developing countries are contributing to an increase in modularization of production and to open innovation processes. There is no single best model of how to compete in this new global environment. However, easier access to global manufacturing capacities and global *explicit/codified* knowledge does necessarily mean that regionally resident capabilities don't play an important role. Because implicit/tacit knowledge does not flow so easily across borders and needs to be nurtured at local level and used to search, absorb and complement access to capabilities located elsewhere, the local environment still plays an important role.

4.3 Local environment and regional competitiveness

There have been a number of studies analyzing the significance of the local environment in the competitive strengths of firms. In the 1920s Alfred Marshall was the first to point out the link between geographical agglomeration and the incidence of external economies, in what he called the 'industrial atmosphere'. Marshall argued that agglomeration of economic activities in a given region created a pool of workers with specialized skills and facilitated the development of specialized inputs and services. Later, geographers such as Perroux [19] also argued that higher-growth propulsive industries would affect other industries via backwards and forwards linkages occurring within a particular geographic space.

Following the Marshallian tradition, other authors in the field of regional economics also put forward a similar argument, proposing the notion of 'industrial districts'. For example Pyke, Becattini and Sengenberger [20], point out the importance of local sharing of common values and beliefs, the sense of belonging and the role that institutions such as family, schools, church, local authorities, local union organisations, and so on, play in the diffusion of such localized common values. According to Michael Storper [21] the 'system of social regulation', including relations of trust, institutional coordination and mechanisms for social consensus, play an important role in coordinating collaboration between enterprises, in fostering the dynamics of local entrepreneurial activity, in the reproduction of labor practices, or more generally in the local dynamics of social reproduction.

Similarly, the concept of 'innovative milieu' put forward by the GREMI group (Groupement de Recherche Européen sur les Milieux Innovateurs) [22–24] stresses the importance of agglomeration of economic and social activities for 'collective learning' and 'uncertainty reduction'. According to Camagni ([24], p.3) an innovative milieu may be defined as "the set or the complex network of mainly informal, social relationships on a limited geographical area, which enhances the local innovative capability through synergetic and collective learning processes". Finally there is also some evidence [25] that geographic proximity between industry and universities facilitates knowledge spillovers from university research to private firms and may lead to higher innovation rates in terms of patenting and R&D expenditure.

In short, a number of studies suggest that the environment close to the firms (physical, economic, social) may be an important factor contributing to their ability to compete in a new global world. This is not just because of the reduction of physical distance and associated transport and location costs, but mainly because it facilitates information exchange, lowers uncertainty, increases the frequency of interpersonal contacts, facilitates trust and the diffusion of common values and beliefs, and promotes collective localized learning. Rosalind Williams [26] argues that "knowledge is global, but learning is local". Local environment conditions play an important role because they foster the so-called nontradable socio-institutional input factors (regional capabilities) that are specific, unique and relatively immobile.

Some production input factors are, however, more mobile than others. For example, capital is much more mobile than land, human capital, or labor. Immobile factors are more important because they cannot be easily transferred and reproduced elsewhere and they represent the regions' unique resource capabilities. This is why companies located in some places have advantages over others by virtue of the dynamics of local learning based upon the appropriate set of local unique resources, such as workers, engineers, managerial talent, suppliers, universities, and so on.

For example, the family-owned firm STIHL based near Stuttgart in Germany, though it could shift production to its lower-wage factories in China and Brazil, often prefers to maintain manufacturing of its most advanced products at home. Similarly, in Portuguese clusters such as moulds, textiles and so on, leading firms often maintain their base manufacturing at home where they can benefit from relatively well trained workers and conditions for achieving better quality.

Another example is the Shan Zhai region in China [27]. The term 'Shan Zhai' today refers to businesses based on fake or pirated products. However, many businesses with Shan Zhai origins are now becoming innovative disrupters and, in many cases, market leaders. Some of them are fast, flexible, innovative, and willing to take risks. As they achieve a certain scale, they quickly move up the value chain and develop core competencies to differentiate themselves from other imitators. The Shan Zhai phenomenon is not about low-cost fake products; it is about how the local environment enables some Chinese companies to achieve global success.

To further illustrate the importance of the local environment, Pisano and Shih ([2], p.3) propose to extend the notion of 'agricultural commons' – in existence in England up to the first industrial revolution – to the notion of 'industrial commons'. While 'agriculture commons' relate to the land and resources shared by a community of farmers, industrial commons are:

> [...] webs of technological know-how, operational capabilities, and specialized skills that are embedded in the workforce, competitors, suppliers, customers, cooperative R&D ventures, and universities, and often support multiple industrial sectors ... needed to turn ideas and inventions into competitive, commercial products.

These technical and organizational capabilities are embedded in local networks of people (workers) and organizations i.e. competitors, suppliers, customers, cooperative R&D, universities, and so on, and they flow across organizations through movements of people from one organization to another, through supplier-customer collaborations, formal and informal technology sharing and imitation of competitors [2]. According to Berger ([18], p.14):

> [...] much learning takes place as companies move their ideas beyond prototypes and demonstration and through the stages of commercialization. Learning takes place as engineers and technicians on the factory floor come back with their problems to the design engineers and struggle together with them to find better resolutions; learning takes place as tacit knowledge is converted into standardized and codified processes; as end-users come back with complaints that need to be fixed.

In our view these processes of learning and knowledge accumulation may follow different trajectories. Jensen et al. [28] propose an interesting dichotomy that contrasts the STI-mode and the DUI-mode of knowledge accumulation.

The Science–Technology Innovation mode (STI-mode) might be referred to as the classic, top-down, internal, research and innovation (R&I) model first practiced in large corporate laboratories, transformed into an externalized model of university laboratory research adapted to technological innovation through 'academic entrepreneurship'. It is the source of start-up and spin-out small and medium enterprises (SMEs) in high-tech clusters, sometimes characterized in terms of a 'patenting – seed/angel/venture fund – incubator' model of new business growth.

This contrasts markedly with the Doing–Using Interacting (DUI-mode) approach to innovation. This is not immediate exploitation of laboratory bench knowledge, although some such knowledge may lie behind the existent state-of-the-art technology or even contribute to its furtherance. DUI involves knowledge recombination among diverse knowledge and practice sets. Accordingly it is fundamentally *interactive* among firms and/or intermediaries characterized by 'related variety' in the first instance. However, research shows that such is the potential of Schumpeterian knowledge recombination that many innovations integrate very different firms/sectors or institutional knowledge-sets.

Accordingly, DUI is *diversified* in that it thrives on cross-fertilisation or cross-pollination of ideas and practices from different fields, for example the intelligent textiles for stay-clean car seats that inspired the innovation of bacteria-free medical uniforms. This means that DUI is *inclusive* for firms that have the needed information about a shared innovation possibility, provided demonstration effort is made. The entailed knowledge for DUI is thus *implicit* rather than codified, and *regional/Local* rather than globally available.

4.4 Reframing the industrialization debate

While in the previous section we argued that the creation and use of unique immobile nontradable socio-institutional local factors associated with different kinds of localized knowledge trajectories and learning is very important for regional competitiveness, in this section we frame the role that local manufacturing capability, particularly in the DUI-mode (but also in STI mode with new possibilities for micro manufacturing), may play in the construction and sustainability of these relatively unique local knowledge resources.

It is not, as often argued [14] that manufacturing is important because of employment creation, or because of multiplier and scalability effects which impact on productivity. The presence of local manufacturing enables the development of specific localized learning linkages between production, R&D and design, hence contributing to the success of innovation. Both Pisano and Shih [2] and Berger [18] argue that manufacturing is one key local capability of any innovation ecosystem, and as such cannot be delocalized without creating 'capability holes'.

Many other studies undertaken in the US such as [29–31], amongst others, also point out the importance of retaining and counteracting the delocalization of manufacturing activities. Apparently, according do Pisano and Shih [2], if a region loses its capacity to manufacture it may also eventually lose its capacity to innovate. Local manufacturing works like an 'anchor' for local learning, dissemination of technology and good management practices.

However, although manufacturing and innovation share the same 'commons', the crucial linkages between product design, prototyping, testing and production that fuel regional innovation, do not work in the same way across all industries. Amongst other things, it depends on the product design choice of modularity. Because of the widespread use of new design and production technologies enabling integration of product and processes, the choice of modular product architectures is today much wider.

The innovation management literature sees modularization as being related to product architecture i.e. components and subsystems as modules and how they connect to each other [32]. However, we also have to look to the rapidly increasing industrial capabilities in developing countries if we want to find out why and how companies combine the choice of product architecture, with strategies to delocalize manufacturing and access remote capabilities [33].

In addition, we need to look beneath the surface of physical components and subproducts. Behind each product or component, we find core competencies grounded in technical and organizational capabilities that enabled their creation, production, and delivery. In addition, these underlying networks of capabilities are dynamic [34] i.e. new capabilities emerge to change the possibilities of both products and processes.

One interesting framework to analyze the extent to which manufacturing is tightly linked to the local innovation system, i.e. whether it does or does not play a key role in learning and innovation is to look at various degrees of modularity and process maturity. Pisano and Shih [2] propose to use the matrix in Figure 4.1

Process maturity: the degree to which the process has evolved	Low	High
High	**Process-embedded innovation** Process technologies, though mature, are still highly integral to product innovation. Subtle changes in process can alter the product's characteristics in unpredictable ways. *Design cannot be separated from manufacturing.* EXAMPLES Craft products, high-end wine, high-end apparel, heat-treated metal fabrication, advanced materials fabrication, specialty chemicals	**Pure product innovation** The processes are mature, and the value of integrating product design with manufacturing is low. *Outsourcing manufacturing makes sense.* EXAMPLES Desktop computers, consumer electronics, active pharmaceutical ingredients, commodity semiconductors
Low	**Process-driven innovation** Major process innovations are evolving rapidly and can have a huge impact on the product. The value of integrating R&D and manufacturing is extremely high. *The risks of separating design and manufacturing are enormous.* EXAMPLES Biotech drugs, nanomaterials, OLED and electrophoretic displays, superminiaturized assembly	**Pure process innovation** Process technology is evolving rapidly but is not intimately connected to product innovation. *While locating product design near manufacturing is not critical, proximity between process R&D and manufacturing is.* EXAMPLES Advanced semiconductors, high-density flexible circuits

Modularity: the degree to which information about product design can be separated from the manufacturing process (Low → High)

Fig. 4.1. Product modularity vs process maturity.

If a product falls into the upper-left quadrant (process-embedded innovation), design tends to be less modular, i.e. product architecture is integrated, and therefore the commons between product design and process manufacturing are much higher. In this situation one cannot easily disentangle R&D and design from production, and therefore the presence of local production capabilities is vital for the regional innovation system. Craft industries, or high-end wine brands (such as Porto and others), specialty apparels (e.g. fire resistant, swimming, etc.) are examples of industries where product modularity is relatively low, manufacturing processes are quite developed and mature and therefore design cannot be separated from manufacturing.

If a product falls into the lower-left quadrant (process-driven innovation) the situation is more or less similar. In sectors such as biotechnology drugs and nanomaterials, product architecture is essentially integrated but manufacturing processes may not have yet reached maturity in order to be safely and efficiently transferred elsewhere. As in the previous case this innovation requires proximity between product design and manufacturing.

When the product architecture is essentially integrated, i.e. with low modularity, the argument that a region can focus on high value services of design and or R&D and let others do the innovation translational work of testing, preseries and ramp-up manufacturing may lead to loss of both. Possibly the region will first lose manufacturing operations where processes are more mature, therefore easier to transfer, but in time it may also lose less developed or less mature manufacturing operations.

In the upper-right quadrant (pure product innovation) products are highly modular and manufacturing processes fully mature. In industries such as consumer electronics or active pharmaceutical ingredients amongst others, it is very difficult to retain manufacturing operations, since they can be easily transferred to cheap labor locations. For regions with relative comparative advantage in labor costs, it may make sense to capture some share of international outsource investment in these areas, perhaps looking for a *near shore* strategy at a relatively close location.

Finally, if a product falls into the lower-right quadrant (pure process technology), product architecture is highly modular but process maturity is low, i.e. manufacturing processes are poorly developed though process technology is evolving rapidly. In these sectors such as advanced semiconductors colocation of product R&D and manufacturing may not be essential, and therefore it makes no sense to compete with regions that may offer lower costs and technologically advanced processes.

As we can see from Figure 4.1, higher modularity may offer multiple opportunities, but, on the other hand, it may also have some negative effects. First it encourages fragmentation of production and second it enhances the possibilities for outsourcing and offshoring design and R&D, since it discourages wide-ranging research, i.e. research on every system and component.

4.5 New challenges for a regional innovation policy agenda

Although everyone today agrees that European regions need higher rates of growth and jobs creation, there is little agreement on what regional and multilevel innovation policies should be implemented. In a world of fragmented GVCs, where firms (not just manufacturing firms, but also banks, law firms, accountants, retailers, and others) may have little or no particular interest or commitment to the development of their local community, what can regional policies do to strengthen local capabilities for innovation?

One typical reaction of European regional policy to this new world of distributed industrial and service capabilities presented earlier appears to be entirely based on a post-industrial view of the world [10]. Because services as a percentage of total economic activity (GDP) tend to increase as economies become more developed, loss of manufacturing is deemed unproblematic and automatically driven by the 'forces of globalisation'.

Also, given the extension across national borders of today's value chains, local linkages and interdependences among domestic firms may be less relevant than extra-regional/national linkages of global scope [35] and therefore the relevance of the resources under the control of public authorities in a given territory would tend to diminish. Regional or national industrial innovation policies are therefore often seen as irrelevant, losing adherence and legitimacy not only in the eyes of business owners but also in the eyes of the citizens in general. In this perspective regional policy should only focus on value-added services, design, creative industries, knowledge intensive start-ups and on spin-offs of academic research. It may also consider facilitating the attraction of hubs and decision centers of large multinational enterprise networks.

Second, another typical policy approach is to argue for more of the same. In European policy documents [36] we find the belief that industrial policy is needed because it creates jobs and because industry creates and retains upstream and downstream interactions with other sectors. This view leads to reinforcement of more of the same 'traditional' policies, aimed at enhancing existing industries or industrial clusters, hence overlooking the degree to which modularity in these industries may eventually lead to delocalization.

We believe all these reactions are partial and often distorted by a misconception of the role of local learning and innovation in international regional competitiveness. As seen in previous sections this new global and more interdependent world is having profound consequences for management of international production and operations in GVCs, for open innovation and for regional innovation policies in a multilevel context. We need to rethink innovation policies in order to better reflect this change and to provide governments with policy tools that effectively help regions to create and protect their capabilities while at the same time being connected to global science networks and to global markets.

Nevertheless, we cannot assume that localized collective learning and construction of regional capabilities is automatically driven just by colocation or agglomeration of complementary or related economic activities. That is, we cannot assume that appropriation of the so-called Marshallian external economies is deemed unproblematic when carried out in physical closeness. Antonelli and Quéré [37], amongst others, argue that it is not sufficient that technological externalities are "freely available in the air" for effective technological communication to take place. Regional innovation policies should therefore focus on stimulating and capturing the benefits arising from local collective learning, innovation and entrepreneurship.

Innovation policies should first focus on creating the base conditions for the development of 'commons'. This is best undertaken at national or European level. Just as the so-called 'smart specialisation strategies' [38] argue that at the European-level policy should concentrate on multipurpose technologies, we favor the idea that for 'commons' to flourish at the European and national levels we need to build a massive portfolio of technological competences. This can only be done by decades of continuous public investment in basic and applied research. The idea is not that this investment leads directly to an increase in product or services design and innovation, but that it leads to the formation of human capital.

In every region, every university and public and semi-public research institutes should seek to participate in large European challenge-driven science projects around general multipurpose science and technology fields in order to create or improve local conditions for 'commons'. Also at European and national levels policies should enlarge their scope beyond science and technology, to include education, information society, health, transports, agriculture, and so on.

Second, at the regional level innovation policies should focus on creating capabilities for translation and transformation of these multipurpose knowledge competencies into innovative products, services or business models through support to testing, pilot productions, speeding up ramp-up operations, testing of production in international operations, and so on.

As we argued earlier, manufacturing and local production plays a vital role in creating and sustaining capabilities for undertaking such translational work from knowledge into goods, services or business models, since it provides key learning linkages between science and innovation. Local governments should therefore be clear about their targets. They are not targeting specific sectors, industries or clusters. Their target is creation of capabilities to transfer knowledge into prototyping, pilot production, customer development and testing.

However it also follows from our earlier discussion based on Pisano and Shih that local policies should only focus on two types of manufacturing capabilities. Those pertaining to immature, or newly emerging, process technologies and those in contexts in which product architecture is less modular and manufacturing-process innovation is therefore highly integrated with product design and R&D. In both cases, regions will find that their competitive advantage in a global world lies in creating the capabilities grounded on the need to have R&D, design and production processes geographically close.

Bibliography

[1] OECD. Interconnected Economies: Benefiting from Global Value Chains – Synthesis Report. OECD, Paris, 2013.
[2] Pisano, G. P., Shih, W. C. Producing Prosperity – Why America needs a manufacturing renaissance. Boston: Harvard Business Review Press, 2012.
[3] Gereffi, G. The organization of buyer-driven global commodity chains: How U.S. retailers shape overseas production networks. In: Gereffi, G., Korzeniewicz, M., eds. Commodity Chains and Global Capitalism, Westport, CT: Praeger, 1994, pp. 95–122.
[4] Feenstra, R., Hamilton, G. Emergent Economies, Divergent Paths: Economic Organization and International Trade in South Korea and Taiwan. Cambridge: Cambridge University Press, 2006.
[5] Sturgeon, T. J. Modular production networks: a new American model of industrial organization. Industrial and Corporate Change 11(3), 2002, 451–496.
[6] Friedman, T. The World is Flat: A Brief History of the Twenty-first Century. Farrar. Strauss and Giroux. 1st edn., April 2005.
[7] Ohmae, K. The Borderless World. New York: Harper Collins, 1990.
[8] OECD. The Internationalisation of Business R&D: Evidence, Impacts and Implications. OECD, Paris, 2008.
[9] Vernon, R. International Investment and International Trade in the Product Cycle. The Quarterly Journal of Economics, MIT Press, United States, 1966.
[10] Bell, D. The Coming of Post-industrial Society: A Venture in Social Forecasting. New York: Basic Books, 1973.
[11] Castells, M. The Rise of the Network Society. Oxford: Blackwell, 1996.
[12] Borrus, M., Ernst, D., Haggard, S., eds. International Production Networks in Asia: rivalry or riches? London: Routledge, 2000.
[13] Barboza, D. Supply Chain for iPhone Highlights Costs in China. From http://www.nytimes.com/2010/07/06/technology/06iphone.html, 2010. Retrieved May 19, 2014.
[14] Bonvillian, W. B. Reinventing American Manufacturing: The Role of Innovation. Innovations: technology, governance, globalization 7(3), Making in America, 2012, 97–125.
[15] Berger, S. How We Compete: What Companies around the World are Doing to Make it Today's Global Economy. Crown Business, 2006.
[16] Dicken, P. Global Shift: The Internationalization of Economic Activity. 2nd edn. Guilford, 1992.
[17] Hayes, R., Pisano, G., Upton D., Wheelwright, S. Operations Strategy and Technology: Pursing the Competitive Edge. John Wiley & Sons, 2005.
[18] Berger, S., ed. Making in America – From innovation to market. Cambridge MA: The MIT Press, 2013.
[19] Perroux, F. Note sur la notion de pôle de croissance. Économie Appliquée 8, 1955, 307–320.
[20] Pyke, F., Becattini, G., Sengenberger, W., eds. Industrial Districts and Inter-Firm Cooperation in Italy. Geneva: International Institute for Labour Studies, 1990.
[21] Storper, M. The resurgence of regional economies 10 years later: the regions as a nexus of untraded interdependencies. European Urban and Regional Studies 2, 1995, 91–221.
[22] Aydalot, P. Milieux Innovateurs en Europe. Paris: Economica, 1986.

[23] Camagni, R. Local milieu, uncertainty and innovation networks: towards a new dynamic theory of economic space. In: Camagni R., ed. Innovation networks: spatial perspectives. GREMI – Groupe de Recherche Européen sur les Milieux Innovateurs. London and New York: Belhaven Press, 1991.
[24] Maillat, D. Territorial Dynamic, Innovative Milieus and Regional Policy. Entrepreneurship and Regional Development 7, 1995, 157–165.
[25] Audretsch, D. B., Feldman, M. Knowledge spillovers and the geography of innovation and production. American Economic Review 86(3), 1996, 630–640.
[26] Williams, R. Retooling: A Historian Confronts Technological Change. Cambridge: MIT Press, 2002.
[27] Johnson, B. Shanzai! Wired Magazine. From http://www.wired.co.uk/magazine/archive/2011/01/features/shanzai, 2010. Retrieved May 19, 2014.
[28] Jensen, M. B., Johnson, B., Lorenz, E., Lundvall, B.-Å. Forms of knowledge and modes of innovation. Research Policy 36(5), 2007, 680–693.
[29] Council on Competitiveness Make?: An American Manufacturing Movement. Report by the Council on Competitiveness and the US Manufacturing Competitiveness Initiative (USMCI), 2011.
[30] Hart, B. Y. D. M., Ezell, S. J., Atkinson, R. D. Why America Needs a National Network for Manufacturing Innovation. Information Technology and Innovation Foundation, December 2012.
[31] Kaushal, B. Y. A., Mayor, T., Riedl, P. Manufacturing's Wake-Up Call. Strategy + Business 64, autumn 2011. From http://www.strategy-business.com/article/11306?pg=all. Retrieved on 22 June 2014.
[32] Henderson, R. M., Clark, K. B. Architectural Innovation: The Reconfiguration of Existing Product Technologies and the Failure of Established Firms. Administrative Science Quarterly, Special Issue: Technology, Organizations and Innovation 35(1), 1990, 9–30.
[33] Salvador, F., Forza, C., Rungtusanatham, M. Product variety, modularity, and component sourcing decisions: Theorizing beyond generic prescriptions. Journal of Operations Management, 20(5), 2002, 549–575.
[34] Teece, D., Pisano, G. The dynamic capabilities of firms: an introduction. Industrial and Corporate Change 3, 1994, 537–556.
[35] Laranja, M., Uyarra, E., Flanagan, K. Policies for science, technology and innovation: Translating rationales into regional policies in a multi-level setting. Research Policy 37, 2008, 823–835.
[36] COM. Communication from the Commission to the European Parliament, the Council, the European Economic and Social Committee and the Committee of the Regions, For a European Industrial Renaissance. Bruxelas, 22.1.2014.
[37] Antonelli, C., Quéré, M. The Governance of Interactive Learning within Innovation System. Urban Studies 39(5–6), 2002, 1051–1063.
[38] Foray, D. Understanding "Smart Specialisation". In: Pontikakis, D., Kyriakou, D., van Bavel, R., eds. Report to the Erawatch. The Question of R&D Specialisation: Perspectives and policy implications. Sevilha: JRC/IPTS, 2009.

Neeta Baporikar

5 Innovation Knowledge Management Nexus

Abstract: Innovation is about helping organizations grow. Growth is often measured in terms of turnover and profit, but can also occur in knowledge, in human experience, and in efficiency and quality. Innovation is the process of making changes to something established by introducing something new. As such, it can be radical or incremental, and it can be applied to products, processes, or services and in any organization. It can happen at all levels in an organization, from management teams to departments and even to the level of the individual. Various factors encourage and drive an organization to innovate. Each of these drivers demands continuous innovation and learning so that the process can be repeated continuously. These drivers also help to create a sense of urgency around the need to create new organizational goals and generate new ideas for meeting these goals. The term *innovation* is often associated with products. However, innovation can also occur in processes that make products, services, or deliver products and services and also include intangible ones. This chapter focuses on innovation in the organizational context, and describes the main concepts behind innovation, what drives the innovation in organizations and tries to understand the nexus between innovation and knowledge management (KM).

5.1 Introduction

A couple of decades ago, when economists forecast the highest earning countries across the globe, many put their money on Japan as the leader, Germany as the runner-up, and the United States in third place for the largest GDPs in the new millennium. But now that we're seven years into the 21st century, it's clear that those economists lost the bet. The United States' GDP is currently $12.3 trillion, exceeding the current GDPs of Japan and Germany by about $8 trillion and $10 trillion, respectively. So what happened to make the United States' output soar above economists' predictions? According to some leading executives and management thinkers, the answer is innovation. In fact, many argue that innovation is the most important driver of macroeconomics today. That's why it is interesting to understand what drives innovation and to discuss innovation, leadership, and the new economy of creativity, knowledge, and invention – and how to focus these amorphous concepts into real business dollars. These insights are relevant to executives from businesses large and small, global and local. Technological innovation is central to the progress of civilizations and economic and societal prosperity, yet most innovations fail. The primary reason most innovations fail is that end users do not adopt them. The reasons for lack

Neeta Baporikar: Ministry of Higher Education, Scientific Research Department, Salalah, Sultanate of Oman. Email: neetajb@rediffmail.com

of adoption are subtle, but often revolve around insufficient knowledge of end users' preferences and requirements (these factors may also not be consciously known to the end users themselves – these may have to be revealed during 'forced choice' situations). Henry Ford was fond of saying that if he had performed market research prior to developing the automobile, the responses he would have received would not have pointed toward the need to develop automobiles but rather toward the invention of faster horses that ate less hay. Improved marketing research per se will probably not lead to higher rates of successful innovation adoption.

Over the past five years innovation has become one of the top priorities for organizations that want to remain competitive in a knowledge/creative economy. Various studies report the importance given by executives to the implementation of innovation management initiatives [1–4]. Even during a time of financial crisis, organizations continue to strongly believe that innovation can be a solution for preparing them to bounce back once the crisis is over [2]. Innovation can be described through a variety of different lenses – application (product, service, process, paradigm or business concept), level (incremental, substantial, or radical), target (consumer, business, or procedure), and so on. Different technologies can be used to support the various phases of the innovation process but very few are fully integrated and provide the features necessary to support the new managerial approaches and models of innovation. Since knowledge is considered to be a catalyst for innovation we believe that a knowledge-enabled system could be of great value in supporting and leveraging the innovation process, which is currently rarely automated and very often not clearly defined.

5.2 Background

5.2.1 KM and Innovation: The Missing Link

Many people, including management experts and consultants, have an inkling of the linkage between knowledge management (KM) linkage and innovation in some way. Some even go further by saying that innovation would justify investment in KM. However when these people are asked how KM would be linked to innovation, one may receive divergent answers. It is hard to expect a unified answer when the notion of innovation is not immediately clear. It is not uncommon to define innovation as a process that produces new product. But, this is a pretty narrow definition of innovation. We can innovate not only on product and service, but also on business process. In fact, to commercialize innovation, organizations need to combine product/service innovation with business process innovation.

The above paragraphs shed some light on the mystery of innovation. However, it is not possible to completely understand innovation without addressing the question of who should innovate. Some experts opine that only those people, who are qualified, such as researchers or senior management, should be involved in the innovation process. Others think that everybody, including the customer and in the case of 'open innovation', the public, should be involved in the innovation process. "Who should be involved in the innovation process?" is a question that can never be answered satisfactorily, because we do not know for sure who has brilliant ideas, the nutrients for innovation, in their mind. Thus, the best thing to do is connect people who have some common ground, but with enough diversity, to spur innovation. To put it simply, we need what is called 'creative abrasion' to spark innovation. This is where knowledge management (KM) and innovation coincide.

KM is about connecting the right people to knowledge sources, which can be either experts or written records. Of course this involves stimulating conversation that matters, i.e. conversation among people who are bound by mutual interest in a knowledge domain, or shared passion, to solve their organization's problem. It is worth noting that although these people are attracted to the same thing, they may have different views or come from different backgrounds In addition, KM could offer a 'shortcut' to innovation by managing what the organization knows, such as past project experience and lessons learned, which could be used as cues to build new ideas. Thus, KM facilitates innovation.

5.2.2 Knowledge Management and Innovation: How Are They Related?

Companies in today's globalized world must innovate to compete. Many successful companies have found that knowledge management strategies and practices are central to ongoing innovation [5–8]; this chapter brings together research regarding knowledge management processes and practices that are found in organizations that are basically innovative firms. The chapter contends that such practices could be employed across a range of firms to enable and enhance the potential for innovation within firms in multiple sectors.

Innovation is widely recognized as a source of a firm's and indeed a nation's wealth creation. Many western countries have been investigating innovation in attempts to understand and hence enhance the likelihood of increasing innovation [9–13]. The contributions of organizational innovation to organizational performance are of long-standing interest [14–16] and in corporate arenas [7, 8, 17]. Thus the focus herein is to find out the role of different factors of knowledge management in bringing about innovation. Moreover, there are different models of knowledge management and innovation. See Table 5.1. Secondary objectives include: to find out the impact of knowledge management activities in enhancing the knowledge assets of the organizations, to explore the factors underlying the success of knowledge transformation, to investigate the role of the knowledge transformation process in bringing about innovation and to find out the determinants of innovation that affect the knowledge transformation process.

Table 5.1. Models of knowledge management and innovation

Different models of knowledge management and innovation
Harnessing knowledge for innovation [23]
A contingency model for knowledge management capability and innovation [19]
Research innovation and knowledge management [28]
Macro process of knowledge management for continuous innovation [29]

Shankar et al. [18] explored the idea that knowledge management creates long term competitive advantage. Ju, Li and Lee [19] developed a strategic contingency model to identify interrelationships among knowledge characteristics, knowledge management strategy, knowledge integration, organizational learning, knowledge management capability and innovation. They explored the idea that knowledge characteristics with higher modularity and explosiveness could enhance organizational learning and knowledge integration. Furthermore, they found that levels of organizational learning, knowledge integration and knowledge management capability have significant impact on a firm's innovation. Drucker [20] stated that in knowledge economies, knowledge is the primary factor of economic development and conventional factors of production like land, labor and capital have not been abolished but these factors have become secondary. In a contingent era, the organizations that utilize their knowledge in an efficient way will be the industry winners through offering the most innovative products. Neilson [21] integrated the knowledge management with the dynamic capabilities approach by demonstrating that dynamic capabilities could be seen as composed of concrete and well-known knowledge management activities.

Majchrazk, Cooper and Neece [22] developed an approach to reuse the knowledge for innovation by making better understanding of the knowledge reuse process when innovation is expected. They pointed out the problems and approach, including the decision to search for others' ideas to reuse, search and evaluate others' ideas to reuse and develop ideas to find out the performance gap which could be filled by the using others' knowledge. Goh [23] developed an integrated management framework to harness knowledge for innovation. He explained why innovation management should not be viewed as mutually independent from knowledge management. Also, he explained the role of knowledge creation and the value of knowledge capital in support of knowledge to bring about innovation. He said that to bring about innovation organizations should support knowledge-centered principles in order to make an efficient role for knowledge creation for innovation.

Leifer, O'Connor and Rice [24] searched out the role and importance of different hubs to bring about radical innovation in mature firms. They proposed different ways to manage radical innovation projects. Malhotra [25] incorporated knowledge management technologies in business processes of organizations. He provided pragmatic understanding of how to integrate knowledge management strategy, technologies and business processes to get better performances. Plesis [26] examined the role of knowledge management to bring about innovation. She found the factors that are important for knowledge management system for bringing about innovation. She also tailored the value proposition by using personal experiences and literature available on knowledge management and innovation. She found that the world's economic growth has changed due to rapid innovation and new technological shift, short product life cycle and increasing rate of new product development. Massa and Testa [27] explained the relationship between benchmarking and innovation through knowledge management. They found that organizations can get the explicit and tacit knowledge from outside the organization through benchmarking. To bring about innovation, organizations have to integrate the explicit and tacit knowledge that is captured and acquired from inside and outside the organization.

5.3 Innovation

Innovation has been variously defined and can be examined from a variety of perspectives, from a broad definition such as "innovation refers to the process of bringing any new, problem solving idea into use" [30], to a more outcome-based approach, where "innovation is the process whereby new ideas are transformed, through economic activity, into sustainable value-creating outcomes" [31]. Innovation in general is used to describe new products, processes and services undertaken by firms, which lead to an increase in performance. The notion here is of a change which leads to a commercial process. Similarly, organizational innovation is defined as the adoption of an idea or behavior that is new to the organization, where "the innovation can be a new product, a new service, a new technology or a new administrative practice" [32]. The changing nature of the market, the challenge of ongoing change and the emergence of the knowledge society has lead to an increased focus on innovation. Innovation is required because we cannot expect that the accumulated competence, skills, knowledge, product services and structure of the present will continue to be adequate [16]. Innovation implies improving on existing products and processes, finding new ways and also abandoning the old, or reviewing every product, service, technology, market, and distribution channel on a regular basis [16]. Much of the research on innovation in firms has come from studies of research and development (R&D) or technological changes. In research and development, the focus on knowledge involves both the creation of knowledge and the reuse of knowledge. Of particular importance in knowledge creation is the notion of sharing knowledge, particularly the tacit knowledge that has not been codified.

These studies have shaped our understandings of innovation and these understandings have since been applied to studies of innovative firms. In both R&D and innovative firms, "knowing what we know" is enhanced by a culture of knowledge sharing, which facilitates the flow and generation of new knowledge, hence capturing and using the internal knowledge base as well as being open to new ideas and technologies from external sources. The enablers of knowledge creation are seen as a knowledge-sharing inclusive culture, organizational structures which encourage interaction, leadership from people in sharing processes and learning networks. The goal is often to facilitate people contacts, such as identifying experienced people who can share their knowledge and providing access to repositories of knowledge. In knowledge practices, the focus is on connecting people. This includes company experts, communities of practice, and enhancing and enabling networks through shared social activities. In addition intraorganizational meetings, technological fairs, roundtables, scientific symposia, or technical and marketing forums can enhance a firm's internal repositories.

There are several definitions of innovation. Harkma [33] stated that the foremost and basic purpose of innovation is to produce new knowledge which can develop and find out the doable solutions for society. Innovation is a practice and process that captures, acquires, manages and diffuses knowledge with the aim of creating new knowledge which will support the production and delivery of distinctive and idiosyncratic kinds of products and services [34]. Plessis [26] delineated innovation as the formation of new knowledge which helps new business returns, which has the purpose of making an organization's internal business process and structure more sophisticated to produce to the market acceptable products and services.

So we can define innovation as: "Activities and processes of creation and implementation of new knowledge in order to produce distinctive products, services and processes to meet the customers' needs and preferences in different ways as well as to make process, structure and technology more sophisticated in such a way that can bring prosperity among individuals, groups and into the entire society."

5.3.1 Importance of Innovation

The basic objective of innovation is to create value for the business. In today's competitive era innovation is the soul of the business because through innovation organizations produce unique products and services. Innovation is also important because of the rapid change in taste and preferences of the customer in emerging and developed markets. That is why according to a research 75 percent of CEOs of the fastest-growing organizations claim that their strongest weapon to compete in market is their innovative products and processes. The organizations which are not very capable of producing innovative products and services will be wiped out from the industry by their competitors because innovation works as a fuel for the organization to grow in any type of environment.

5.3.2 Scope of Innovation

Innovation might be radical or incremental. A radical innovation is a product, service and process with entirely unique or significant improvements in existing features which improve the cost and performance. Radical innovation is highly risky for the business because radically innovated products are more difficult to commercialize. On the other hand, radical innovation in product, service or process is crucial for the business because it involves the development and application of new technology. An important aspect of radical innovation is to what extent new technology is more sophisticated and advanced compared to current technology [35, 36]. Another idea is of different hubs to bring radical innovation. Among those hubs, one important hub is of idea generators. Idea generators are responsible for generating distinctive ideas and there are people who exploit these distinctive ideas, idea hunters, who actually exploit and execute these ideas. Idea gatherers are basically receivers of the ideas. They have skills, expertise, judgment and motivation to respond to these unique ideas. The combination of generators, hunters, and gatherers plays an important role in bringing radical innovation about in large organizations. There are two reasons that firms strive to bring about radical innovation. First, these radical innovations create barriers for the potential competitors and ruin the market share of existing industry players. Second, competitors are capable of developing or producing radically innovated products [37]. Plessis [26] explained that incremental innovation is basically a modification in a product, also called 'line extension' or 'market pull innovation'. Incremental innovation does not need to significantly diversify from current business. That is why this type of innovation enhances the skills and competencies of the organizational employees. Incremental innovation is decisive for the organization because it helps the organization to increase their market share and to remain in industry for a long time.

5.3.3 Innovation at the Firm Level

In the last decade, studies of innovation at the firm level have identified common components in innovative organizations [8]. These components include strategic approaches, linkages and high involvement of staff.

Distinctive features of each component are summarized in Table 5.2.

Similar findings from other recent research on innovative firms found underlying capacities for innovation include vision and strategy, a competency base, creativity and idea management, organizational intelligence, organization and process, culture and climate [7]. These findings confirm our understanding that the ability of an organization to recognize the potential of an innovation is not a simple process and "is a function of how it collects and processes information, depends on nature of the innovation, the organization structure, systems, people, local environment and man-

Table 5.2. Components of the Innovative Organization. Source: Tidd et al., [8] 1997: 314.

Component	Key features
Vision, leadership and the will to innovate	Clearly articulated and clear sense of purpose stretching strategic intent; 'top management commitment'
Appropriate structure	Organization design that enables high levels of creativity
Key individuals	Promoters, champions, gatekeepers and other roles which energize or facilitate innovation
Effective team working	Appropriate use of teams to solve problems. Requires investment in team selection and building
Continuing and stretching individual development	Long term commitment to education and training to ensure high levels of competence and the skills to learn effectively
Extensive communication	Within and between the organization and outside. Internally in three directions – upwards, downwards and laterally.
High involvement in innovation	Participation in organization-wide continuous improvement activity
Customer focus	Internal and external customer orientation. Total quality culture
Creative climate	Positive approach to creative ideas, supported by relevant rewards system – a 'winner's culture'
Learning organization	Processes, structures and cultures which help institutionalize individual learning. Knowledge management

agerial dominant logic" [38]. In summary, innovation cannot be directly created. Indeed, successful innovation in an organization is based on strategy, is dependent on both effective internal and external linkages, usually requires enabling mechanisms to make change happen, and only happens within a supporting organizational context [8]. These views of requirements for innovative firms can also be applied to firms with successful knowledge management strategies and practices. For example organizations need a strategy to manage their knowledge [39, 40], internal and external linkages [41, 42], enabling mechanisms for change [43]. A more market oriented approach to innovation suggests that successful innovations are characterized by the correct anticipation of customer needs, detailed knowledge of the supply chains, and intelligent application of external technology. "All resources of a company, internal and external have to be integrated" [5]. This view contends that the benefits gained through knowledge management in innovative firms include not only the identification of technical competencies that are key to success (technical core competencies), but also the ability to communicate technical core competencies throughout the management of the whole company, with particular focus on R&D and marketing. In addition, the identification of technical core competencies with an opportunity to focus on issues of protection, exploitation and enhancement of competencies is essential.

It is clear from these descriptions of successful innovation that the 'management' of knowledge is central to the organization's innovation. We proceed to further discuss these concepts and use this overlap to identify knowledge management practices that contribute to innovation in more detail.

5.4 Knowledge Management

KM is an organizational process that aims to create a centralized knowledge source within the organization that acquires, assimilates, distributes, integrates, shares, retrieves and reuses the internal and external, explicit and tacit knowledge to bring about innovation in the organization in the form of the product, people and organizational process. Polyani [44] first identified the duality of the knowledge. He divided knowledge into two types:

1. Tacit knowledge
2. Explicit knowledge

5.4.1 Tacit Knowledge

Polyani [44] defined tacit knowledge as abilities, expertise and conceptual thinking. Further, he argued that tacit knowledge is not only attributed to what is known, but it is also attributed to the knower as well. This is because the knower's knowledge level is sometimes soaring but he cannot explain in efficient way, or the knower sometimes does not have adequate sources to disseminate his knowledge to the person who actually needs it. Tacit knowledge is very difficult to acquire because it is embedded in the form of capabilities, skills and ideas which individuals carry in their minds. Tacit knowledge can only be seen through the application; that is why tacit knowledge is difficult to capture, exploit and diffuse among the organizational members.

5.4.2 Explicit Knowledge

Polyani [44] said that explicit knowledge can be disseminated and shared in the form of hard data, well defined procedures, and standardized principles. Nonaka, Takeuchi [45] defined explicit knowledge as "Knowledge of Rationality". Explicit knowledge is easy to capture, manage, share and disseminate to the people.

5.4.3 Relationship Between KM and Innovation

Messa and Testa [27] stated that organizations must develop the receptors that gain or absorb the external knowledge and this activity is strongly correlated with innovation capability. Further, they said that through benchmarking, organizations can acquire explicit and tacit knowledge from external sources. These external sources of knowledge can be integrated with the organizational internal explicit and tacit knowledge and if a knowledge gap prevails, that can be filled through the new knowledge acquisition, which will be helpful in bringing about innovation. Ju et al. [19] argued that in order to get competitive advantage organizations should continuously learn from outside sources. Through the proper knowledge distribution and sharing, organizations can bring about innovation. So, organizations must develop such channels within the organizations through which employees share their knowledge with one another. Plessis [26] stated that innovation depends upon knowledge. So, to bring about innovation, organizations must identify knowledge capability and richness.

Parlby and Taylor [46] asserted that the foremost purpose of knowledge management is to bring about innovation. Plessis [26] stated that organizations can develop collaborations across the organizational boundaries to bring about innovation and to get a sustainable competitive advantage. This collaboration helps the organization to approach new knowledge that can be helpful to fill knowledge gaps within the organization. This collaboration ultimately brings about innovation in the organization and this collaboration can reduce the risk and cost to bringing about innovation.

Organizations that rapidly capture and implement new knowledge across the organization are better able to foster innovation as compared to those organizations that don't focus on this aspect [47]. Furthermore, they argued that the first and most important aspect of innovation is to increase the innovation capability to identify and capture the tacit knowledge of the organization. Tacit knowledge can be acquired from outside the organization such as through customers, suppliers, bankers and so on. This acquisition of tacit knowledge plays a significant role in fostering the process of innovation. Tacit knowledge becomes more important in those particular industries where explicit knowledge is scarce. Through knowledge management, organizations can identify their tacit knowledge that they usually do not know before. Knowledge management also helps the organization to articulate tacit knowledge in the form of explicit knowledge and this is a strong base to bring innovation.

Knowledge management integrates different types of tacit and explicit knowledge. Through integration, organizations can discover what type of tacit and explicit knowledge subsists in the organization. Furthermore, knowledge activities like knowledge gathering, managing, sharing, learning, reuse and retrieval play important roles in bringing about innovation. Through knowledge management activities, organizations find out the extent of knowledge from inside and outside their organizations. Organizations manage this knowledge in the form of datsbases, so that, they can ensure the availability of the right type knowledge for the right person at the right time.

5.5 Conceptual Framework

The basic aim of this research is to explore the integrated approach to knowledge management and other factors which play an important role in bringing about innovation in any industry which was lacking it before. There are several components involved in this model such as KM activities, knowledge transformation, technology and culture and more importantly organizational knowledge assets which are comprised of two things: 1) human capital, and 2) data warehouses. All organizations have both tacit and explicit types of knowledge. Explicit knowledge is easy to disseminate and share with the people whereas tacit knowledge is very difficult to share, integrate and disseminate to the people. The one tool to increase knowledge and enhance the innovation process is open communication within the organization because open communication and flexible structures urge people to create new ideas and share their tacit knowledge.

5.5.1 Information and Communications Technology (ICT) Factors

As far as relationship between IT and KM is concerned, there are two schools of thought. MecDermott and O'Dell [48] stated that KM could be successful without IT. It can be used when it is necessary. On the other hand, some have argued that IT has become much more important because of globalization. It is true that technology alone cannot play any role to capture, manage and exploit the knowledge which exists inside and outside the organizations. Rather, it is the combination of technology and human capital that leverages the KM activities [49]. Today, IT supports the most important tasks of KM. ICT identifies and gathers the knowledge through different tools like web portals, internet, intranet, and so on. Not only does knowledge gathering modernize IT tools but it also helps organizations to diffuse the explicit and tacit knowledge.

ICT plays an important role in organizational communication. Technology is a tool to support the communication. There must be a proper communication structure through which people can share knowledge and ideas. One of the major sources for gaining new knowledge is the internet. Mohammad, Stonkosky and Murray [50] stated that the real challenge for IT experts is to revolutionize the objectives to select, develop and implement better technology that could serve the KM in an efficient and effective way.

5.5.2 Knowledge Management Activities

Neilson [21] made the connection between different knowledge management activities like knowledge creation, acquisition, capturing, assembling, sharing, integration, leverage and exploitation. He further divided these eight KM activities into three dynamic capabilities such as knowledge development, knowledge (re)combination and knowledge use. These eight KM Activities contain all the important activities that start from acquiring new knowledge and end at the exploitation of new knowledge. Through these activities, organizations find out new knowledge within the organization as well as from outside the organization that enhances the knowledge capability of the organization. These knowledge management activities enrich the organizational knowledge assets. This consistent acquisition of new knowledge makes more sophisticated organization processes and routines. And, by the application and use of this new knowledge, innovation can be brought into the organization.

5.5.3 Knowledge Assets

Organizational knowledge assets are the soul of innovation as knowledge assets increase the knowledge capability of the organization and knowledge capability leads to innovation. Knowledge assets include two factors: human capital and knowledge repositories.

5.5.4 Human Capital

The most important factor of organizational knowledge is human capital. Knowledge and competencies of workers have become the vital component of developed economies [51]. Human capital is the most sustainable, inimitable source of competitive advantage. Human capital consists of competencies, skills, knowledge and information possessed by the workers of the organization. Human capital creates the ideas that are the strongest base of innovation. The role of human capital is not only limited to idea generation; rather its role is also important during distinctive idea execution to deliver the innovative products and services to the customers. Organizations must know the type and level of the organizational human capital. In highly dynamic and competitive environments, the collective knowledge and expertise of employees must be utilized in an effective and efficient way so that they could create optimum economic benefits. It is not only important to discover competent human capital in the organization but organizations must create the environment to utilize the competent human capital. Thus, if organizations have a supportive culture and environment, such human capital will produce more economic value.

5.5.5 Knowledge Repositories

Organizational knowledge repositories consist of large databases, data warehouses, internet, intranet, and so on. Knowledge repositories have complete databases of skills, expertise and knowledge of organizational employees. Through knowledge repositories it becomes easy to access information and the knowledge of the organization. Through knowledge repositories, we can accomplish the important task of managing the explicit knowledge of the organization. This explicit knowledge can be diffused to the person or area where it needed.

5.6 Knowledge Transformation Success and Innovation

In this chapter we have discussed several factors that contribute to bringing about innovation, but these factors alone cannot bring about innovation. Rather organizations have to make the knowledge transformation/conversion process successful so that these factors may effectively bring about innovation. Cumming and Teng [52] identified several factors that play an important role in the transfer of knowledge among individuals, groups and in entire organizations. They divided these key factors that make knowledge transfer successful into four broad contextual domains: knowledge context, relational context, recipient context and activity context.

5.6.1 Knowledge Embeddedness

Knowledge used to be entrenched in individuals, tools, processes and in related activities and networks of organizations. One way to transfer knowledge from one place or department to another place or department is to transfer knowledge individuals. The benefit of this is that an organization can transfer both types of knowledge, tacit and explicit, at the same time [53–55]. Whenever there is difference of knowledge and expertise between knowledge recipient and knowledge sender, the knowledge recipient fails to learn. Knowledge can be rooted in organizational activities and practices [56]. Finally, knowledge can also be embedded in multiple elements and sub-networks [51]. It is difficult to transfer knowledge within the organization without mobility of experts with recognized patterns of working mutually. Knowledge embeddedness negatively and significantly affects knowledge transfer success [51].

5.6.2 Knowledge Articulability

Knowledge articulability is an important factor that affects knowledge transfer success. Knowledge articulability is to what extent knowledge is written down, verbalized and articulated [56]. Polyani [44] stated that people explain less than what they know since individuals have tacit knowledge which is unarticulated, intuitive and can only be observed through application. Articulated knowledge can easily be captured, stored and shared with other employees because it can easily be understood and observed as there is less ambiguity in articulated knowledge. Knowledge tacitness is significantly and positively correlated with ambiguity [57]. Transfer of knowledge success depends upon how much knowledge is tacit and explicit. If knowledge is more explicit and articulated the chances of transfer success will be more. Ambiguous and less articulated knowledge is difficult to share and poorly articulated knowledge is difficult to diffuse among organizational employees.

5.6.3 Organizational Distance

The base of organizational distance is the means through which the source and receiver share the knowledge. Organizations can get knowledge from within the organization as well as outside the organization. Knowledge transfer within the organization is easier as compared to outside the organization. Knowledge is easy to transfer from selected parties [58]. For example, franchises [59], chains [60] and networks [61] can transfer knowledge more effectively and efficiently. Thus it can be inferred that transfer success will decrease with the increase in organizational distance between source and recipient of knowledge.

5.6.4 Knowledge Distance

Knowledge distance is to what extent source and recipient have the same kind of knowledge. It has been found that the organizational learning knowledge gap between source and recipient should not be too great [63]. The reason behind this is that if the knowledge gap is greater the learning steps will also be larger and knowledge transfer will be much complex, difficult and time consuming. Hence, it can be said if the knowledge and expertise gap among source and recipient is great, transfer of knowledge and learning will almost be impossible. To minimize these gaps, there must be some adjustments in knowledge and other factors in order to make the knowledge transfer successful. Cumming and Teng [52] found that there is a significant and negative relationship between knowledge distance and the success of knowledge transfer.

5.6.5 Physical Distance

Physical distance is the difficulty, type and expense of getting face to face communication and for knowledge transfer. Athanassiou and Nigh [63] found that face to face interactions are better compared to all other modes of delivering strategically important matters. Moreover, they found that physical distance negatively affects the success of knowledge transfer.

5.6.6 Project Priority

Different projects take different degrees of attention, resources and time. When the recipient gives too much priority to the project, he will be very motivated to get new information and knowledge transfer success with rapidity and with much more intensity. Researchers have identified different variables in making knowledge transfer successful, such as as motivation and learning intent of the recipient of knowledge. These factors play a vital role in knowledge transfer success. Accordingly, people will support the transfer of knowledge in highly prioritized projects than less prioritized ones.

5.6.7 Learning Culture

Learning culture is also an important factor for the success of knowledge transfer. Knowledge transfer has two aspects; first is knowledge velocity and second is knowledge viscosity. Knowledge velocity is the speed of knowledge transfer and knowledge viscosity is the richness of knowledge transfer [64]. When there are learning routines in organizations, every employee starts to get new knowledge by interacting with other people and also by approaching other means such as books, journals, and so on.

5.7 Knowledge Transformation Process

Modes of knowledge transformation as by Nonaka et al. [45] are:

1. Socialization (tacit to tacit)
- Wandering inside
- Wandering outside
- Tacit knowledge transfer
- Tacit knowledge accumulation

2. Externalization (tacit to explicit)
- Dialogue
- Metaphor

3. Combination (explicit to explicit)
- Collecting data and acquisition
- Disseminating data and information
- Editing and synthesizing data and information

4. Internalization (explicit to tacit)
- Personal experience
- Simulation

An organization's top management must focus on this knowledge transformation process because it is the most important source of diffusion of knowledge among individuals, groups and in the entire organization from the top level to the bottom level in the hierarchy of the organization. This knowledge transformation process creates the leverage within the organization regarding knowledge sharing, creation, dissemination and integration of the knowledge. Important tools for knowledge transformation might be mentoring, coaching formal and informal meetings and seminars and it also includes learning by doing. Through these different activities, erudite persons share their knowledge and expertise with others, which can boost the level of knowledge in persons lacking knowledge as well as urging people to gain new knowledge and ideas and to produce something distinctive as compared to competitors which is the soul of innovation and competitive advantage. The biggest achievement for any organization is when their employees start to think in different ways, when they are passionate, devoted and motivated to push their organization to the height of success and excellence. The knowledge transformation process is affected by many factors as well.

5.8 Knowledge Transformation, Collaboration and Integration for Innovation

Knowledge collaboration is very important for bringing innovation. Collaboration might be internal or external. Through internal collaboration, organizations come to know about diverse knowledge that exists in the organization in the form of tacit and explicit knowledge. Through strong internal collaboration with the employees, organizations come to know what, where and how much knowledge exists in the organization. This internal collaboration can foster innovation because when organizations collaborate and integrate with the internal employees this can lead the organization towards the generation of a pool of expertise and creativity which are essentials for bringing about innovation. As far as the role of external collaboration to bring about innovation is concerned, external organizations play an important role, such as customers [65] and competitors [66]. Nowadays, to bring about innovation, it is crucial to make linkages with external organizations to get the knowledge and capabilities that are necessary for innovation [67, 68]. To bring about innovation, linking with external organizations and partners is a core process. Therefore it can be argued that if any firm wants to innovate successfully in a highly complex environment, the innovation process must be supported by an open and flexible structure. Organizations must integrate external knowledge and capabilities. This integration and collaboration allows the organization to get more innovative ideas which is the soul of innovation.

5.8.1 Organizational Learning and Innovation

To compete in a highly dynamic environment each and every organization must focus continuously on learning because customers' needs and preferences are rapidly changing and to meet those requirements organization must seek and learn in new ways. These new ways and methods can only come into the organization through learning. Organizational learning enhances the organization's knowledge capability and knowledge assets. Consequently, organizational learning strengthens the knowledge transformation process because when employees learn they have to share their experiences and knowledge with others who really need current knowledge to fill the knowledge gap. Wijenhoven [69], states that organizational learning urges people to enhance the organizational knowledge base. Organizational learning enhances the interaction among the employees so that knowledge sharing, integration and dissemination are achieved. It will not only boost the quality, and quantity of the information and accumulation of knowledge in a dynamic environment, but also enhances the ability to create new knowledge and its application.

5.8.2 Organizational Culture and Innovation

Culture is one of the most important factors to implement in the knowledge management system. Delong and Fahey [43] stated that knowledge management faces difficulties in being implemented from corporate culture, and that's why organizations normally do not get the maximum benefit from knowledge management. In a study of 453 firms, more than half of them indicated organizational culture was the biggest obstacle to implementing the KM system in the organizations [70]. In order to implement the knowledge management system effectively, organizations must create the thirst for knowledge and achievement among the individuals of the organization. So, to implement knowledge management, organizations need to build the knowledge culture within the organization in which new knowledge acquisition and sharing will be the integral part of the strategy and culture. Gold et al. [71] stated that an encouraging and supportive culture will help to build the knowledge management system in the organizations.

5.9 Innovation Knowledge Management Nexus

The intertwining relationship between innovation and knowledge is seen in the approaches to increasing knowledge and its transfer in organizations. For example, Davenport and Prusak suggest that the best way an organization can transfer knowledge effectively is to "hire smart people and let them talk to each other" [72]. A similar prescription is provided for innovation – the ingredients are clever people and the processes are the interaction, context and a culture of knowledge sharing, or finding effective ways to let people talk and listen to one another. Some writers describe the importance of the generation of knowledge in more detail. Skyrme [42] describes two processes. First, knowing what you know i.e. having better awareness, sharing and application of existing knowledge including that which originates outside the organization. Second, faster and better innovation i.e. more effective conversion of ideas into products and processes [42]. He also describes innovation as a set of interacting knowledge processes. These processes include the absorption of existing knowledge from the external environment, the creation of new knowledge through creative thinking and interchange of ideas, the rapid diffusion of ideas and insights through knowledge networking; the validation, refining and managing of innovation knowledge, matching of creative ideas to unmet customer needs and in solved problems, and encapsulating and codifying knowledge into an appropriate form such as a tangible product, a production of a new internal process, training material for a new service a marketable design, patent, and so on [42]. Each of these sources of knowledge present different challenges and often require different knowledge processes. These forms of knowledge and the processes to implement them are summarized in Table 5.3.

The contribution of knowledge to the continued success of companies such as 3M, Hewlett Packard and GlaxoWellcome is well known [42]. Benefits of knowledge include the avoidance of costly mistakes, such as the sharing of best practices in Chevron, and faster problem solving through videoconferencing by BPAmoco where offshore oil platforms can tap into expertise elsewhere. Other examples include faster development times through learning networks and linking customer problems to an ideas database, and better customer solutions.

A well known example is the sales and support staff at Buckman Labs who use their knowledge repository K'Netix to gain access to the best expertise and to develop innovative solutions to tricky customer problems. Other benefits obtained from knowledge management practices include gaining new business, improved customer service and reduction of risk.

Innovative success in small and medium sized firms is determined by the presence of organizational, technological and marketing competencies and if these competencies are jointly present, firms are more likely to innovate successfully [73]. This study did not specify knowledge practices per se, but the combination of a strong knowledge base, proactive management of innovations and management of the relationship with the environment were major factors in the success of these firms.

Table 5.3. Forms of Knowledge Processes. Source: Developed from Skyrme (1999) [42].

Forms of knowledge	Knowledge Processes
Customer knowledge	Developing deep knowledge through customer relationships, and using it to enhance customer success through improved products and services
Knowledge in products and services	Embedding knowledge in products and surrounding them with knowledge intensive services
Knowledge in people	Developing human competencies and nurturing an innovative culture where learning is valued and knowledge is shared
Knowledge in processes	Embedding knowledge into business processes, and giving access to expertise at critical points
Organizational memory	Recording existing experience for future use, both in the form of explicit knowledge repositories and developing pointers to expertise
Knowledge in relationships	Improving knowledge flows across boundaries: with suppliers, customers and employees, and so on
Knowledge assets	Measuring intellectual capital and managing its development and exploitation

R&D firms and innovative firms have similar and different processes and practices for creating and using knowledge for innovation.

5.9.1 KM Processes in Innovative Firms

The description of the underlying capacities for innovation includes vision and strategy, a competency base, creativity and idea management, organization and process, culture and climate and intelligence [7]. A firm can of course be at different levels with respect to different innovative capacities. Little's study grouped multiple notions of knowledge and knowledge management under the heading of organizational intelligence. This notion described knowledge management as the generation, protection and stewardship of technology and technological knowledge, acquisition and development of knowledge from outside or absorptive capacity. Also included were knowledge articulation and deployment, awareness of own performance and limitations, commitment to understanding the customer's both current and future (unarticulated) needs, structured thinking about the future, scanning the horizon, recognition, screening and selection of new ideas and understanding and using networks for intelligence. Managing a firm's knowledge assets is crucial in innovation and Leonard

Barton [74] focuses on three types of skills or knowledge: public or scientific, industry specific and firm specific.

Other studies have investigated knowledge management practices in leading companies in terms of knowledge flows [75] and found key enablers of knowledge in culture, infrastructure and IT tools and standards.

On the other hand, an overview of successful innovation in largely international firms argues that "knowledge creation is completely dependent on individuals, on human beings" and that "people are the primary source of innovation in high performance organizations. Both personnel co-location and job rotation help to build a network of informal linkages that provide information channels that are seen as often far superior and efficient than formal reporting structures and official message boards. In particular, mutual trust and face to face contacts are to be maintained in order to sustain the network" [5].

Table 5.4. Knowledge – innovation processes. Source: Self developed.

Knowledge Processes	Activities	Contexts
Making external connections as routine Sharing explicit and tacit knowledge Build networks and gain tacit knowledge Broader access to explicit knowledge	Visiting scientists programs Informal meetings and conferences Job rotation Multimedia technology/ICT	Informal links and networks
Learn from new approaches/technologies Diversity of experiences and challenges High visibility – new ideas diffused Common knowledge captured/accessible	Cross-functional teams Intercultural project managers High impact projects Shared databases	Projects and processes
Deep expertise Disciplinary ideas/multiple approaches Diffusion of technology	Functional specialists Multifunctional prototyping Dual career ladders Technology Agents	Hierarchical and functional
Connections for sharing tacit knowledge Expertise from top with local knowledge Increase potential for learning from diverse situations and challenges	Face to face meetings Expatriates Local recruiting International dispatchments	Regional and local

5.9.2 Knowledge Processes and Practices for Innovation

Taking into account the contextual nature of knowledge, Boutellier et al. [5] have suggested a number of activities that assist in knowledge sharing in a number of contexts, including informal links and networks, and projects and processes. Underlying these activities are specific knowledge processes that can be encouraged. The knowledge processes inherent in these activities and contexts have been added to the focus on knowledge creation processes in the first column in Table 5.4.

Such processes include the sharing of tacit and explicit knowledge, interactive processes, building networks, increased knowledge from diversity of ideas and experience, and diffusion of knowledge.

5.10 Solutions and Recommendations

5.10.1 Guidelines for Motivating Innovation

- Entrust firefighting. One cannot drive innovation when one is putting out operational fires. It is better to hire the best operations team one can, and then stay out of their way.
- Establish credibility. Trust breeds innovation, and communication breeds trust. Establish a formal communication program.
- Acknowledge criticism. Not every idea is a good one, and some are downright lousy. To improve ideas, ruthlessly seek out criticism.
- Attest it. Does the idea save money, increase real productivity, and equally important, is it feasible? Nothing ruins credibility faster than a business case full of holes. Better to do good homework and get some feedback before shopping the idea around.
- Gaze around. Staying inside your organization and keeping the lights on may be instinctual during down times, but it is hardly a pathway to innovation. Look outside the frame of reference.
- Keep away from technology reverence. A project need not involve brand new technologies to be innovative.
- Re-examine the startups. In addition to innovating inside their own companies, CEOs have a role to play in driving innovation in the industry they belong in.
- Constrict expenditure. Constraint breeds innovation, as it's very tempting, when money and resources flow freely, to stick with tried and true solutions. When money and resources are constrained, one has to find new and creative ways to solve problems.
- Timing. It's all in the timing, as expecting operating units to participate in a new project at the drop of a hat is a surefire formula for failure.
- Discover opportunities in problems.

5.11 Future Research Directions

Innovation and KM-related issues will be critical for organizations in this millennium. For organizations to ensure quality products and services there is a need to do research into the knowledge management issues of innovation and its process. ICT-enabled KM will make it all the more difficult for organizations to really grasp and apply innovation strategy, hence understanding innovation strategy implementation would also provide insights. Case study analysis of innovative approaches and successful KM implementation may be undertaken on a longitudinal basis and the cultural aspects both organizational and individual also need to be deeply studied and rationally interpreted. Further work differentiating knowledge practices from central components of innovation in firms is another interesting and understudied area yet to be undertaken.

5.12 Conclusion

In this chapter, an integrated model that includes several factors that play vital roles in bringing about innovation is developed for holistic understanding. ICT factors help the organization to find out and manage organizational knowledge, which increases the organizational knowledge assets and capabilities. Knowledge success factors are vital in diffusing knowledge from individuals to the entire organization, which strengthens the organizational knowledge culture. In order to speed up the innovation process, organizations must implement the innovation determinants that are actually the cause of innovation. This chapter brings together literature from research on innovation, both specifically from R&D contexts and from research on innovative firms, and the importance of innovation for success at the firm and national level has been demonstrated. The essential contributions from knowledge practices and their critical role in innovative firms have been identified. The chapter concludes with some ways in which the activities seen as central to innovation in firms encapsulate knowledge management practices.

In today's fast-paced marketplace, if a company keeps offering the same product, a rival can easily race past with a better one. And yet another competitor will blow them both out of the water when it invents something altogether different and better – something innovative. To remain competitive, companies must consider how to find and keep visionary leaders and how to foster innovation and creativity in their employees, the executives and experts at the event agreed. On the global stage, innovation could mean the difference between the United States keeping a tight grasp on economic leadership or eventually slipping behind countries like China and India, as some economists have predicted. However, those fast-growing countries also face the same challenge. Thus, innovation and KM is critical not only to business success, but to its very survival. Innovation is so important because fresh approaches, new ways of delivering ideas, visual and content changes all keep people interested and coming back. Innovation is not exclusive to big brands, but is, in fact, essential for all brands. Hence, aligning innovation strategy with KM strategy and then to overall business strategy is critical.

Any organization needs a method to capture ideas and provide a readily available source of basic concepts to feed into the corporate innovation process. But it needs more than that. Innovation is essentially a creative activity and requires freedom of thought and action. To accommodate this, a formal KM framework must be established to channel the activities into corporate assets. And the earlier this journey begins the better for organizations in today's highly competitive, networked knowledge economies.

5.13 Key Terms and Definitions

Creativity: the ability to transcend traditional ideas, rules, patterns, relationships, or the like, and to create meaningful new ideas, forms, methods, interpretations, and so on. It includes originality, progressiveness, or imagination.
Innovation: something new or different introduced, it is the act of innovating that includes introduction of new things or methods.
Knowledge: acquaintance with facts, truths, or principles, as from study or investigation, familiarity or conversance, as with a particular subject or branch of learning and includes acquaintance or familiarity gained by sight, experience, or report. It is the fact or state of knowing, the perception of fact or truth by clear and certain mental apprehension.
Learning: knowledge acquired by systematic study in any field of scholarly application. It also is the act or process of acquiring knowledge or skill.
Organization: a group of persons organized for some end or work; an organized structure or whole for a business or administrative concern united and constructed for a particular end or a body of administrative officials, as of a political party, a government department, and so on.
Process: a systematic series of actions directed to some end; it is a continuous action, operation, or series of changes taking place in a definite manner. A process is thus a series of progressive and interdependent steps by which an end is attained.

Bibliography

[1] AMA. The Quest for Innovation. A global study of innovation management 2006–2016, 2006.
[2] Andrew, J. P., Haanaes, K., Michael, D. C., Sirkin, H. L., Taylor, A. Innovation 2009. Making hard decisions in the downturn. Boston, MA, USA: The Boston Consulting Group, 2009.
[3] Capgemini. Global CIO Survey. The Role of IT function in Business Innovation. Innovator vs. Operator: Capgeminiconsultingo, 2008.
[4] IBM. Expanding the Innovation Horizon – The Global CEO Study 2006.
[5] Boutellier, R., Gassman, O., von Zedtwitz, M. *Managing Global Innovation: Uncovering the Secrets of Future Competitiveness*. Berlin: Springer-Verlag, 1999.
[6] David, P., Foray, D. Economic Fundamentals of the Knowledge Society. *International Social Sciences Journal*, Special Issue 171(February/March), 2002.

[7] Little, A. D. *The innovative company: Using policy to promote the development of capacities for innovation*. Final Report to participating governments. Cambridge, UK, 2001. Available on http://www.environmentrisk.com/articles/pdf/innovative.pdf. Accessed on 10 May 2014.

[8] Tidd, J., Bessant, J., Pavitt, K. *Managing Innovation: Integrating Technological, Market and Organizational Change*. Chichester, UK: John Wiley and Sons, 1997.

[9] *Backing Australia's Ability*, Innovation statement, 29 January 2001.

[10] OECD. *Boosting Innovation: The Cluster Approach*. Paris, 1999a.

[11] OECD. *Managing National Innovation Systems*. Paris, 1999b.

[12] OECD. *Mobilizing Human Resources for Innovation*. Paris, 2000.

[13] OECD. *Innovative Networks: Cooperation in National Innovation Systems*. Paris, 2001.

[14] Burns, T., Stalker, G. M. *The Management of Innovation*. London: Oxford University Press, 1961.

[15] Burns, T., Stalker, G. M. *The Management of Innovation*. Oxford: Oxford University Press, 1994.

[16] Drucker, P. *Managing for the Future*. Oxford, England: Butterworth Heinemann, 1992.

[17] Hamel, G. *Leading the revolution*. Boston: Harvard Business School Press, 2000.

[18] Shankar, R., Gupta, A. 'Towards Framework for Knowledge Management Implementation'. Knowledge and Process Management 12(4), 2005, 259–277.

[19] JU, T. L., Li, C. Y., Lee, T. S. "A Contingency Model for Knowledge Management Capability and Innovation". *Industrial Management and Data System* 106(6) (2006) 855–877.

[20] Drucker, P. "The Theory of Business". *Harvard Business Review* Sep/Oct (1994) 95–104.

[21] Nielson, A. P. "Understanding dynamic capabilities through knowledge management". *Journal of Knowledge Management* 10(4) (2006) 59–71.

[22] Majchrzak, A., Cooper, L. P., Neece, O. E. "Knowledge Reuse for Innovation". *Journal of Knowledge Management Sciences* 55(2) (2004) 174–188.

[23] Goh, A. L. S. "Harnessing Knowledge for Innovation: An Integrated Management Framework". *Journal of Knowledge Management* 9(4) (2005) 6–18.

[24] Leifer, R., O'Connor, G. C., Rice, M. "Implementing Radical Innovation in Mature Firms: The Role Hubs". *The Academy Of Management Executive* 15(3) (2001) 102–113.

[25] Malhotra, Y. "Integrated Knowledge Management Technologies in Organizational Business Process: Getting Real time Enterprise to Deliver Real Business Performance". *Journal of Knowledge Management* 9(1) (2005) 7–28.

[26] Plessis, M. D. "The Role Of Knowledge Management in Innovation". *Journal of Knowledge Management* 11(4) (2007) 20–29.

[27] Messa, S., Testa, S. "Innovation or Imitation? Benchmarking: A Knowledge Management Process to Innovate Services". *Benchmarking: An International Journal* 11(6) (2004) 610–620.

[28] Diem, Diem, Ho. "Research, Innovation and Knowledge Management: the ICT Factor". UNESCO, 2007.

[29] Xu, J., Houssin, R., Caillaud, E., Gardoni, M. Macro process of knowledge management for continuous innovation. Journal of Knowledge Management 14(4) (2010) 573–591.

[30] Kanter, R. M. *The change masters: Corporate entrepreneurs at work*. London: Routledge, 1983.

[31] Livingstone, C. Innovation Lecture, Warren Centre. Sydney, April 2000.

[32] Hage, J. T. Organizational Innovation and Organizational Change, *Annual Review of Sociology* 25 (1999) 597–622.

[33] Harkama, S. "A complex perspective on Learning within Innovation Projects". The *Learning Organization* 10(6) (2003) 340–346.

[34] Gloet, M., Terziovski, M. "Exploring the Relationship between Knowledge Management Practices and Innovation Performances". *Journal of Manufacturing Technology Management* 15(5) (2004) 402–409.

[35] Govindarajan, V., Kopalle, P. K. Can incumbents introduce radical and disruptive innovations?: Report No. 04–100. *MSI reports: working paper series*, 2004.

[36] Christensen, C. M., Overdorf, M. "Meeting the Challenge of Disruptive Change." Harvard Business Review 78(2) (2000) 66–76.
[37] Christensen, C. M. The innovator's dilemma: when new technologies cause great firms to fail. Cambridge: Harvard Business Press, 1997.
[38] Afuah, A. *Innovation management: Strategies, implementation and profits*. New York: Oxford University Press, 1998.
[39] Hansen, M., Nohria, N. & Tierney, T. What's your strategy for managing knowledge, *Harvard Business Review* (March/April 1999) 106–116.
[40] Zack, M. H. Developing a Knowledge Strategy, *California Management Review*, 41(3) (Spring 1999) 125–145.
[41] Hansen, M. T. The Search Transfer Problem: the Role of Weak Ties in Sharing Knowledge across Organizational Subunits, *Administrative Science Quarterly* 44 (1999) 82–111.
[42] Skyrme, D. J. *Knowledge Networking: Creating the Collaborative Enterprise*. Butterworth Heinemann, Oxford University Press, 1999.
[43] Delong, D. W., Fahey, L. "Diagnosing Cultural Barriers to Knowledge Management". *Academy of Management Executive* 14(4) (2000) 113–127.
[44] Polyani, M. "Personal Knowledge: Towards a Post Critical Philosophy". *University Of Chicago Press, Chicago, IL*, 1962.
[45] Nonaka, I., Takeuchi, H. "The Knowledge-Creating Company". Oxford: *Oxford University Press*, 1995.
[46] Parlby, D., Taylor, R. "The Power of Knowledge: A Business Guide to Knowledge Management". available at www.kpmgconsulting.com/index.html, 2000.
[47] Cavusgil, S. T., Calantone, R. J., Zhoa, Y. "Tacit Knowledge Transfer and Firm Innovation Capability". *Journal of Business and Industrial Marketing* 18(1) (2003) 6–21.
[48] Mecdermott, R., O'Dell, C. "Overcoming Cultural barriers to sharing knowledge". *Journal of Knowledge Management* 5(1) (2001) 76–85.
[49] Medermott, R. "Why Information Technology inspire but cannot deliver Knowledge Management". *California Management Review* 41(4) (1999) 103–117.
[50] Mohammed, M., Stankosky, M., Murrey, A. "Knowledge Management and Information Technology: Can they work in perfect harmony". *Journal of Knowledge Management* 10(3) (2006) 103–116.
[51] Drucker, P. F. "Post Capitalist Society". Oxford: *Butterworth Heinemann*, 1993.
[52] Cumming, J. L., Teng, B. S. "Transferring R&D Knowledge: They Key factors affecting knowledge transfer success". *Journal of Engineering and Technology Management* 20 (2003) 39–68.
[53] Allen, T. J. "Managing the flow of technology: Technology Transfer and dissemination of technological information within R&D organization". Cambridge MA: *MIT Press*, 1977.
[54] Berry, D. C., Boardbent, D. E. "The Relationship between Task Performance and associate verbalized knowledge". *The Quarterly Journal of experimental psychology* 36A (1984) 209–231.
[55] Starbuck, W. H. "Learning by Knowledge intensive firms". *Journal of Management Studies* 29 (1992) 713–738.
[56] Szulanski, G. "Exploring Interstickness: Impediments to the transfer of best practices within firm". *Strategic Management Journal* 17 summer special issue (1996) 27–43.
[57] Bressman, H., Dirkinshaw, J. M. "Knowledge transfer in International acquisition". *Journal of Internal Business Studies* 30(3) (1999) 439–462.
[58] Simonin, B. L. "Transfer of Marketing Know How in International Strategic Alliances: An Empirical Investigation of the role of antecedents of knowledge ambiguity". *Journal of International Business Studies* 30 (1999) 463–490.
[59] Darr, E., Argot, L. D. "The acquisition, transfer and depreciation of knowledge in service organization: Productivity in franchises". *Journal of Management Sciences* 41 (1995) 1750–1762.

[60] Baum, J. A. C., Ingram, P. "Survival-Enhancing learning in the Manhattan Hotel Industry". *Journal of Management Sciences* 44 (1998) 996–1016.
[61] Uzzi, B. "Sources and Consequences for the Economics Performance of Organizations". *American Sociological Review* 61 (1996) 674–698.
[62] Hamel, G. "Competition for Competence and Inter-partner Learning within International Strategic Alliances". *Strategic Management Journal* 12 (1991) 83–103.
[63] Athanassiou, N., Nigh, D. "Internationalization, Tacit Knowledge and Top Management Teams of MNCs". *Journal of International Business Studies* 31(3) (2000) 471–487.
[64] Davenport, T., Prusak, L. *Working Knowledge: How Organizations Manage What They Know.* Boston MA: Harvard Business School Press, 1998.
[65] Gassman, O., Sandmeier, P., Wecht, C. H. "Extreme Customer Innovation in the Front End Learning from a New Software Paradigm". *International Journal of Technology Management* 33(1) (2006) 46–66.
[66] Hamel, G., Doz, Y. L., Parahald, C. K. "Collaborate with your Competitors and Win". *Harvard Business Review* 67(1) (1989) 133–139.
[67] Chesbough, H. "Open Innovation: The Imperative for Creating and Profiting from Technology". *Harvard Business School Press*, Boston 2003.
[68] Powell, W. W., Kopat, K. W., Smithdoerr, L. "Inter-Organizational Collaboration and the Locus of Innovation: Networks of Learning in Bio-Technology". *Administrative Science Quarterly* 41(1) (1996) 116–145.
[69] Wijenhoven, F. "Acquiring Organizational Learning, Norms: A Contingency Approach for Understanding Deutero Learning". *Management Learning* 32(2) (2001) 181–200.
[70] Ruggles, R. "The State of the Notion: Knowledge Management in Practice". *California Management Review* 40(3) (1998) 80–89.
[71] Gold, A. H., Malhotra, A., Segars, A. H. "Knowledge Management: An Original Capability Perspective". *Journal of Management Information System* 18(1) (2001) 185–214.
[72] Davenport, T. H., Prusak, L. "Working Knowledge, How Organizations Manage What They Know". Boston MA: *Harvard Business Process*, 1998.
[73] Cobbenhagen, J. *Successful Innovation: Towards a New Theory for the Management of Small and Medium sized Enterprises*, New Horizons in the Economics of Innovation Series. Cheltenham UK: Edward Elgar, 2000.
[74] Leonard Barton, D. *Wellsprings of Knowledge: Building and Sustaining the Sources of Innovation.* Boston MA: Harvard Business School Press, 1995.
[75] Armbrecht, F. M. R., Chapas, R. B., Chappelow, C. C., Farris, G. F., Friga, P. N., Hartz, C. A., McIlvaine, M. E., Postle, S. R., Whitwell, G. E. "Knowledge Management in Research and Development". *Research Technology Management*, (July/August 2001) 28–48.

Filomena Antunes Brás

6 Human Capital Accounting: A Contribution to Innovation Management or a Fairy Tale?

Abstract: Since the 1970s human capital accounting (HCA) has been a promising contribution to management, namely from a management control perspective. Most managers hold very positive attitudes towards HCA, but the integration of HCA in the management control process has never really been attained, much less in financial accounting and reporting. In this chapter, the role played by HCA within organizational performance is discussed, namely where it concerns measuring, reporting and controlling human capital.

6.1 Introduction

Management practices have undergone many innovations. Companies have been downsized, delayered, and hollowed out. Newly trained and empowered employees have implemented many innovative practices including continuous improvement, reengineering, just-in-time manufacturing, and total quality management [1]. Many of these innovations have fundamentally changed the relationship between the organization and its employees, customers, suppliers, and other stakeholders [1]. In many instances, arms-length transactions between independent parties have been replaced by long-term partnerships in which intangibles such as service, innovation, and flexibility are essential to success. Intangible and difficult-to-measure resources are driving the creation of wealth in many companies [1]. That is the reason why many authors have been advocating that organizations increasingly rely on intangible assets for competitive advantage. Human capital is an ever more important intangible asset because it is the basis for the development of all the other intangibles. Yet, managing this intangible, when financial accounting sees it as an expense in the income statement, remains a challenge.

It was a challenge back in the 1970s when the problem was first identified. But almost fifty years later we are not even close to solving the problem. Human capital, intellectual capital, knowledge management, and innovation management are several terms to identify mainly the same concept whilst tackling the problems of being valued, measured and also reported. As Flamholtz, Bullen and Hua [2] stated, most of the relatively easy preliminary research is accomplished, that is the identification of the problem and development of cost and value models to measure human capital. However, the remaining research required to develop this field of study is com-

Filomena Antunes Brás: Department of Management, School of Economics and Management, University of Minho. Email: filomena@eeg.uminho.pt

plex, there are not many scholars interested in performing this work, and it requires the cooperation of organizations willing to serve as research sites for applied research studies [2]. Moreover, we believe that this field of research needs contributions from several disciplines to develop and achieve the expected results.

Surprisingly, or maybe not, there are still calls for the need to measure human capital in order to manage it, since this capital is crucial in today's developed societies. Case studies and cross-sectional analyses have provided evidence that intangibles are the fundamental source of competitive advantage for business companies in most industrial sectors. Therefore, the successful management of companies requires that intangibles must be identified, measured and controlled [3] because companies based on material assets are no longer able to achieve further economies of scale, and they are unable to gain competitive advantage with tangible assets alone [3]. Innovation is one possible approach to developing competitive advantage and human capital is inherent to the innovation process.

Human capital pertains to knowledge resources such as innovation, education, training, learning and development, employee demographics, industrial relations, compensation and remuneration, career planning and development, senior executive performance and results, involvement in the community, knowledge identification, sharing and retention of work-related qualifications [4]. People own human capital. People are part of the intellectual capital. People are responsible for maintaining and growing intellectual assets. Their unique contribution comprises their command of information and previous experience, their ability to integrate and use judgment, to be innovative and intuitive, and to develop and use human relationships. This contribution is knowledge, and it is seen as a continual creation in which its flow around the organization is seen as the vital dynamic of progress [5].

The constant generation of new knowledge and experience, and its availability for the benefit of all who can use it effectively, are fundamental issues [5]. Arguably, the most important of all business processes are those designed to accomplish these tasks, and the role of employee development takes on a specifically strategic mantle as a primary means of achieving the goals [5]. When senior management describes 'people' as a strategic asset, they describe employee performance and behaviors that help execute the company's strategy. But just as organizational performance is a function of people and systems, the appropriate human resource system is required to select, develop and reward employees in ways that produce strategic behaviors [6].

Although measuring human resources/human capital is a challenge to accountants, from the human resource management (HRM) perspective it is also rather important because measuring human resources' performance is an increasingly important concern for human resource professionals, senior managements and CEOs. The challenge here is to answer the CEO or senior management team when they ask the human resource function to justify its contribution to the organization [6]. Therefore, some authors argue that managing the human resource function efficiently is important. However, they warn that focusing on cost reduction as the primary measure of human resource's performance will ultimately result in human resources being managed like a commodity, rather than a strategic asset [6]. The human resource function contributes to company performance by driving human capital to activities that, due to strategic positioning, means performing different activities from rivals or performing similar activities in different ways [6].

Since the 1970s, human capital accounting (HCA) has been promising its contribution to management, namely from a management control point of view. Most managers hold very positive attitudes towards HCA, but the integration of HCA in the management control process has never really been attained. In this chapter, the role played by HCA within organizational performance is thoroughly discussed in several areas of study.

This chapter is structured in the following manner. First we present and discuss HCA from the accounting literature point of view. Then, we present and discuss the call for developing HCA by the strategic and measurement branch of human resources management (HRM) literature, followed by the implications of HCA for the emergence of the intellectual capital field. Stakeholders want to know how human capital is managed and to what extent human capital management practices are successful. Therefore, we present the reporting problem associated with HCA and the challenges that assist the development of this field of research. We end by drawing conclusions on the state of the art of HCA.

6.2 The Evolution of Human Capital Accounting

We need to go back to the 1960s and 70s to fully understand the emergence of human capital accounting. The designation of this field of study has also evolved over time. In the 60s and 70s, the field was identified as human resource costing and accounting (HRCA). With the emergence of intellectual capital literature and the balanced scorecard, some authors named it human capital accounting (HCA). Today, there are also authors who designate this field as intellectual capital accounting, due to the fact that human capital is a component of intellectual capital. In this chapter we use these designations indiscriminately because they all embrace the human capital concept, and they address the tracking, measuring and controlling of this capital. In sum, these different designations represent the evolution of thought which has occurred in the management and accounting sciences.

Brummet, Flamholtz and Pyle's paper [7] is one of the first works in the area of HCA. They addressed the issue of how to put people in the balance sheet. They noticed what managers were saying, and are still saying today, in corporate annual reports: "our employees are our most important – and most valuable – asset". However, if we look closer into the remainder of the corporate annual report, we might ask what information we can get from this 'asset' – "where is this human asset on these statements which serves as reports on the company's resources and earnings? What is the value of this 'most important' or 'most valuable' asset? Is it increasing, decreasing, or remaining unchanged? What return, if any, is the firm earning on its human assets? Is the firm allocating its human assets in the most profitable way?" [7]. The information was (and in general still is) absent from the corporate annual report.

At this stage, several cost and value-based models were developed. For example, Hekimian and Jones [8] proposed the concepts of original cost, reposition cost and opportunity cost related to human resources; Brummet, Flamholtz and Pyle [7] proposed a human resources accounting system for investments in managers; Hermanson [9] proposed two methods for valuing human resources in the balance sheet, using discounted earnings and wages approaches (the unpurchased goodwill method and the adjusted-present value method); Flamholtz is the most preeminent author in the field and proposed several models, namely the model for measuring human resource positional replacement cost, and the model of the determinants of an individual's value to a formal organization [10]; and, Lev and Schwartz [11] and Friedman and Lev [12] proposed models to value human resources using economic theory (discounted wage flows).

This initial stage of evolution of this field of research is not free of criticism. In fact, putting people on the balance sheet was heavily criticized because human resources (HR) are not a commodity. Although several HCA cost and value-based models have been developed to tackle these issues, there are some flaws: lack of reliability due to subjectivity and considerable use of estimates are inherent to those models; the company does not actually own human assets because human capital is owned by the HR (people) and, therefore, it is not a commodity; and there is always the risk of employee turnover, that is, the owner of human capital leaves the company. Consequently, normative accounting theory has not been sensitive in considering human capital as an asset, since the international accounting standard of intangibles (IAS 38) does not recognize human capital as an intangible asset. This absence raises another problem. When information on human capital is not mandatory, it is not possible to compare the use and value of this capital across companies. What can we conclude on firms' human capital management performance? It depends on the management narrative, whenever it exists.

However, human capital is not totally absent from the financial statements. It is considered as an expense in the Income Statement. And this has not changed, yet. It has a pernicious effect on the management-making decision process because it is a cost to be set against revenue. In this light profit can be increased by reducing human capital cost. Therefore, from an agency theory perspective, management holds an incentive to make people work harder for the same payment [13], or worse. This has been the management's argument to move industries to 'low cost' economies. Similarly, in the management accounting literature, investments in people such as training, upgrading skills, funding educational courses or redeployment expenses are designated discretionary costs. These are invariably represented as being amongst the most susceptible to reduction in times of declining financial fortunes [13], thus compromising not only the human resources short and long-term interests but also the company's interests in building knowledge stocks capable of providing innovation.

The first stage of the field's evolution had few important implications for external financial reporting. It led to a second stage characterized by a decrease of interest in the field. However, in the 1990s, the increasing importance of concepts such as intellectual capital and intangibles has renewed international interest in HCA theory and practice. The gradual yet fundamental transformation that most of the world's advanced economies made in shifting from industrial economies in which plant and equipment are the core assets, to post-industrial economies in which human capital, knowledge and intellectual property are the core assets, has stimulated interest in the contribution of HCA [2]. The potential success of an organization now lies upon its intellectual capabilities rather than upon its physical assets. Accordingly, organizations must pay attention to the development and deployment of intellectual capital, which includes human capital [2, 5]. The emergence of technology companies has put HCA in the spotlight, because these companies rely more heavily on their human resources/capital than industrial companies [2, 14]. Unfortunately, the accounting discipline has not yet responded to this change, and therefore the critics say that it is likely that investors are paying a price due to lack of information about managerial and human capital [2], in the sense that there are several empirical research studies which determined that HCA has an impact on decision making. For example, Elias' [15] experiment found that external user's decisions on investments in common stock were made differently when resorting to the use of HRCA information. Following this author, Hendricks [16] found that stock investment decisions were significantly affected by additional HRCA information. Schawn [17] further extended Elias' and Hendricks' studies by examining the effects of HRCA information on financial decisions in comparison to decisions based on conventional financial information. Results showed that companies using HRCA information were considered better prepared and the inclusion of human resource accounting information resulted in statistically significantly better predictions of companies' net income [2]. More recently, Hansson [18] and Moreno-Campos [19] reached the same conclusions.

As a result, the measurement tools available cause anomalies, because they are based mostly on costs. But organizations now need systems that continually assess and reassess the people they employ, including their skills, talents and behavioral attributes, while paying attention to how human resources impact the bottom line. HCA is an accounting tool that is relevant to the measurement and, in turn, to the management of intellectual capital and innovation, specifically human capital [2].

HCA has even greater importance as a powerful managerial tool in internal HRM decisions [2]. Flamholtz contributed to the HCA literature with several models, thereby incorporating the monetary and behavioral emphases of human resources valuation. He introduced an alternative perspective on accounting for people. His objective was to demonstrate the value of human resources to businesses (through the model of the determinants of an individual's value to a formal organization) rather than their valuation per se [13]. Once the value of people was appraised, Flamholtz believed that senior management could be persuaded to make better use of their human capital. He also believed that it was possible to develop better accounting practices that would serve the interests of all stakeholders [13]. According to Roslender and Stevenson [13], he faced two difficulties, however. First, at his time any development in management accounting practices was under the influence of the financial accounting principles and reporting. As a consequence, any new approach to accounting for people was destined to be linked to the prevailing thinking of "putting people on the balance sheet". Secondly, and most importantly, in the 1970s "very few within the realms of senior management were persuaded of the necessity to contemplate whether 'our people are our greatest asset'" [13]. Although Flamholtz [10] argued that it was more appropriate to embed accounting for people within the traditions of managerial accounting rather than financial accounting and reporting, he hoped the intellectual capital movement would lead to changes in financial accounting and reporting.

Flamholtz [10] has identified three objectives for HCA: to develop methods for measuring human resource cost and value in order to provide a quantitative basis for decision making by managers and investors; to develop methods of measuring human resource cost and value necessary to monitor the effectiveness of management's utilization of human resources; and to develop a theory explaining the nature and determinants of the value of people to formal organizations. This last objective is more critical – people are a scarce resource that managers must manage as efficiently and effectively as possible in order to ensure that it delivers the greatest benefits to the company [3].

Many stakeholders, even those outside Sweden where HCA was more flourishing, agree on the need for a better transparency of investments in human capital. According to Johanson [20], the Organisation for Economic Cooperation and Development (OECD) position was that improvement on the information and decision making systems that shape human capital acquisition and utilization was a key factor in helping a nation's companies to be competitive with others. The implication is that human capital measurement and accounting for human resources have to be further improved [20]. The topic of HCA was also introduced by the European Commission at the end of 1995 in a white paper on education and learning. The intention of the Commission was to prepare guidelines for action to promote teaching and learning in the member states. One of five general objectives of this work was to "treat capital investment and investment in training on an equal basis" [20].

In sum, despite the pronounced interest in HRCA in Sweden, Turner cited by Johanson [20] remarked that human resource accounting "has progressed at something less than a snail's pace", and it may still be accurate to some extent. In comparison with what has been expected of HRCA, progress has moved at a dilatory pace [20]. Why? We present some evidence in the following sections that might explain this situation.

6.3 Call for HCA from Human Resources Management Literature

"How much does human resource management matter?" continues to be an important question for both practitioners and academic researchers in the HRM field. One of the most effective arguments demonstrating the importance of human resource decisions lies in demonstrating their empirical relationships with valued company outcomes, particularly financial performance. Although both practitioners and researchers have reported substantial money value benefits associated with the use of certain human resources practices, the task of ensuring the validity of these substantive estimates rests most heavily with researchers [21].

According to Becker and Huselid [6], a common source of human resources' measurement problem is that human resources largely represent a cost to be minimized (due to financial accounting that induces this behavior), to adopt performance measures that are limited by efficiency goals. In the absence of a business case for human resources' strategic impact, human resources managers are trapped [6]. In this case, human resources responded by significantly improving its performance on efficiency measures. However, the consequences for the company's performance may not be anticipated or even acceptable, because the 'commoditization' of human resources lead to a losing proposition between accessibility and appropriateness [6]. According to Becker and Huselid, by executing business strategy, human resources share the responsibility for business problems with all managers. Measuring human resources' strategic performance is not about doing the same things human resources professional have always done. It is about developing new measures based on the unique demands of the company's strategy [6].

Many authors from the HRM field, namely Gerhart, Wright and McMahan [21], Huselid and Becker [6, 21], and Huselid, Becker and Ulrich [23] argue that a company's human resources (people) constitute a potential source of major sustained competitive advantage, and it is through human resources practices that companies can leverage the value of people in ways that result in positive performance outcomes. Therefore, management needs to develop an information system that allows them to monitor and link human resources policies to organizational performance. However, most companies use formal performance measurement systems that are extensions of their financial reporting systems [1]. They justify this practice because the financial reporting system provides measures that, on one hand, are generally regarded as reliable and consistent, thereby giving a solid foundation for developing reward and accountability structures; and, on the other hand, it engages with the primary objective of creating profits for owners, thereby giving a performance measurement focus consistent with organizational objectives [1]. However, criticisms of conventional performance measurement systems have been increasing.

Critics argue that financial performance measures lack the requisite variety to give decision makers the range of information they need to manage processes [1], because a critical question is to know whether a company's human capital is growing or not, and to have a more internal and detailed track of this than merely watching how the market values the company's intellectual capital. However, financial accounting does not give an answer to this since the accounting standard does not recognize all the intellectual assets as intangibles assets. Nevertheless, it is a fact of organizational life that "numbers speak louder than words" [5], and as some CEOs say, if it can be visualized it can be measured and if it can be measured it can be managed. Quantification is what is important [5]. Replacement costs of knowledge, systems or people can be calculated [5]. Levels of competence and expertise can be tracked [4]. Calculate financial effects of poor intellectual capital management – for example, the loss of key people, repeated mistakes due to lack of knowledge transfer, lost revenues due to inadequate capability, and so on – can be computed. However, these are outputs. We more commonly track inputs – ratios of investment in training, research and development, information technologies – but such measures alone give us no clues as to how effectively such investments are being used. The convinced proponent of the principles of intellectual capital will always be in a disadvantaged position compared to the financial community, without a set of credible measures for those components that are strategically important [4].

HCA has been developed in three main branches: as a valuation/measurement of human capital (providing numerical information about the cost and value of people as organizational resources); serving as an analytical framework to facilitate decision making; and, as motivating decision makers to adopt a human resource perspective [2].

One role of HCA measurement is to provide numerical information as an input to management and financial decisions. But another and even more important role comes from the measurement process, from the act of monitoring and quantifying the costs and value of people from a human resource perspective [2]. From a managerial perspective, the process of measuring, as well as the measurements themselves, send the message that people are valuable organizational resources and should therefore be managed as such. In the past, the impact of managing layoffs and downsizing could be better planned and monitored if costs involving these operations were tracked and followed in order to be controlled, namely by identifying and predicting the hidden costs associated with these events (e.g. in the layoff situation, the costs associated with rehiring qualified human resources, and the hidden costs associated with impacts on the morale and productivity, and the retention of the people that were not laid off). Nowadays, the focus goes to managing innovation, through a better management of organizational knowledge. HCA technology allows the analysis of the effects of such decisions and a better understanding of the long-term implications and hidden costs of management's business decisions.

However, there are also other motivations for developing HCA. Corporate managers make expenditures which they justify as investments in human resources, but accounting recognizes these investments as expenditures without considering the timing of expected benefits. Therefore, HCA must develop an information system that provides information on how those investments contribute to a company's performance. In other words, by reporting on actions taken and results achieved in relation to objectives and goals, namely bring individuals to the level of technical competence and familiarity required for a given position at the company; by identifying the competences available at the organization, and the ones identified as necessary to achieve organization's goals; and, by making company aware of the replacement costs of individuals or functions. Flamholtz, Searfoss and Coff [24] report on a study taken place at the Touche Ross & Co., one of the 'Big 8' international CPA firms that represented the state-of-the-art of HCA at that time, where the goal of the system developed was concerned with replacement cost of the audit senior position, a key position for company's operations. HRCA was implemented and tested because management was particularly interested in the replacement cost and how its management would enhance HRM.

However, we do not find many case studies in the literature of the practice of human resource accounting. According to Cantrell, Benton, Laudal and Thomas [25], many companies are concerned with recruiting and training top talent but they fail to control for their human resources investments because managers lack the tools they need to accurately measure the return on investment in human capital. Johanson [20] reports that all the firms he surveyed that were adopting some kind of HRCA tools or were starting to use some kind of those tools, managers produced an 'aha' reaction. The use of such tools made them see a clear connection between HRM and the company's financial results.

Cantrell, Benton, Laudal and Thomas [25] tested a tool known as the human capital development framework over more than 60 organizations. Their empirical results suggest that financial performance improves as a company improves its scoring in the critical human capital processes with strong relationships with financial success. Lundberg and Wiklund cited by Johanson [20] found that 70 percent of the personnel managers in Stockholm-based companies with more than 200 employees claimed that they were applying HRCA to some extent. Most of the organizations had started to do so in the beginning of the 1990s. In an investigation conducted by the Swedish Association of Local Authorities in 1994, it was found that 22 percent of the responding 276 Swedish local authorities had decided to use an accounting approach to HRCA. Only 5–15 percent of personnel, accounting and financial managers have declared their lack of interest in HRCA [20]. Sweden seems to be an exception in the sense that HRCA applications are commonly applied [26]. In turn, a survey taken by Accenture [25] revealed that, in fact, many companies do not even make the attempt to measure human capital. Why do companies avoid implementing and taking advantage of having such a measurement system?

6.4 The Emergence of Intellectual Capital – New Demands Over an Old Problem

Nowadays the success of a company focuses more upon its intellectual and systems capabilities than on its physical assets. The capacity to manage human intellect, and to convert it into useful products and services, is becoming the critical executive skill of the age. As a result, there has been a flurry of interest in intellectual capital, creativity, innovation, and the learning organization [27].

One of the first definitions of intellectual capital is that one provided by Edvinsson and Malone [28]. To these authors, intellectual capital is the possession of knowledge, applied experience, organizational technology, customer relationships and professional skills that provide the company with a competitive edge in the market.

In an innovative company the professional intellect (knowledge) is crucial. The professional intellect of an organization operates on four levels [27]: cognitive knowledge (know-what), achieved through training and certification; advanced skills (know-how), the ability to apply the rules of a discipline to complex real-world problems is the most widespread value-creating professional skill level; systems understanding (know-why) is deep knowledge of the web of cause-and-effect relationships underlying a discipline (it permits professionals to move beyond the execution of tasks to solve larger and more complex problems); self-motivated creativity (care-why) consists of will, motivation, and adaptability for success. Highly motivated and creative groups often outperform groups with greater physical or financial resources.

HCA can contribute positively to the innovation management process. Firstly, it can identify the human capital available at the company in terms of competences and knowledge owned by the people working in the company. It can compare the stock of competences and knowledge to the company's needs, in order to define recruitment policies to catch and/or define training programmes to develop those competences and knowledge. It can also help identify the functional positions that are most critical for organizational success. Then, it can associate the costs that underline the reposition of that human capital function, to provide financial information on how it would cost the company to substitute the human capital lost when human resource leaves. Finally, competences and function performed by people are then linked to organizational goals. People's motivation is fundamental to outperform their tasks and goals. A survey can help the company to monitor the HR motivation level in order to not only improve the work environment but also to identify the informal processes that need to be developed to stimulate and nurture human capital's creativity.

Nevertheless, to be successful, the company must first define its strategy and define the guide map that allows it to achieve its strategic goals. In the strategic HRM perspective, there is a call to link strategy to policies of HRM in order to potentiate the organizational performance. The HR Scorecard of Becker, Huselid, and Ulrich [23] is a tool that allows analysis of the impact of HRM policies upon the organizational performance, and therefore to identify what needs to be changed. This information system is available through HCA. However, organizations have made little progress in developing measures of how people (or human resources/human capital) make a strategic contribution [23]. But, managing intangibles assets, like human capital, demands a different accounting system because the traditional accounting system fails to track or analyze these assets. Management tends to overlook their importance and skew their decisions in favor of tangibles when they do not have available reliable figures [29].

Many companies had been making investment decisions in their people largely on the basis of faith rather than empirical, quantifiable data, and instead of resorting to the use of HCA information systems. But as the company moves to instill a more disciplined culture, management knows that facts supported by data would be more valuable than intuition in many instances [24, 27].

Cantrell, Benton, Laudal and Thomas [25] describe a case study of SAP America, a technological company, that was largely unsatisfied with many of the data-based measurement options it had tried, such as economic value-added and return on invested capital because they did not reflect the increasing importance of people assets. According to the authors, the company had also tried other metrics, people-oriented metrics such as training budget per employee, but they proved not to be helpful for making investment decisions, because these numbers did not tell anything about the effectiveness or the company's people programs. On the other hand, survey data collected from employees was seen as helpful but limited in utility. Then, the company tested the human capital development framework [25].

The architecture of the framework comprises four levels. At the first level, the company assesses its human capital processes, such as competency management, rewards and recognition, career development, employee relations, performance appraisal, workplace design and planning, learning and knowledge management, human capital strategy, succession planning, recruiting, and human capital infrastructure. These are the processes that contribute to developing human capital capabilities, the second level of the framework. At this level, the company analyses its human capital capabilities, such as human capital efficiency, talent management, leadership capability, workforce performance, employee engagement, workforce adaptability, and the ability to change. Improved human capital capabilities are the support for driving improvement in key performance drivers like innovation, customer satisfaction, or quality (third level). Finally, key performance drivers are those targets that management might hope to improve financial organizational performance measured through the business results that company aims to achieve, such as total return to shareholders or revenue growth [25]. For each element of the first three levels of the model, it was provided with effectiveness scores on a scale of 1 (low effectiveness) through 5 (high effectiveness).

According to the authors [25], using the information provided in the initial assessment, SAP America had access to better information to help determine where it should focus its human capital investments in order to produce the greatest business benefits. Management based investment decisions on three criteria: the relative effectiveness of each human capital process, the presence or absence of evidence that a given process was strongly associated with financial performance, and the strategic or cultural importance of each process based on its linkage to particular key performance drivers of human capital capabilities. This framework helped SAP America to assess and visualize the importance of each element of the framework. For example, certain key performance drivers like innovation or quality could be of greater strategic importance as compared to others. Likewise, an organization's culture may favor certain human capital capabilities over others. In general, companies will achieve the greatest benefits if they focus on developing processes that are the least mature and most related to either financial performance or the performance of an important key business driver or capability [25]. Cantrell, Benton, Laudal and Thomas [25] showed that in the case they analyzed, SAP America, the ability to deliver a steady stream of innovative new software applications to the customers and to tailor the company to their targeted market segments was vital to its strategy. Hence, management needs to be focused on those processes that can help the company to maintain its innovative capability.

In summary, Cantrell, Benton, Laudal and Thomas' [25] findings suggest that organizations with more mature human capital processes have better financial performance than those organizations with less mature processes. Specifically, they found that those organizations that focus on processes devoted to three key areas – creating a people strategy aligned with the business strategy, providing supportive work environments, and developing employees by giving them ample opportunities to learn and grow (investments in human capital) – achieve far greater economic success than those that do not. Cantrell, Benton, Laudal and Thomas [25] found statistically significant relationships between an organization's innovation capability and employee engagement; employee engagement in turn has statistically significant relationships with almost all the elements of the human capital processes (10 in 13). These results suggest that organizations competing on innovation will achieve better financial results by improving employee engagement through improving a broad range of human capital processes.

Other case studies and cross-sectional analyses have provided evidence that intangibles are the fundamental source of competitive advantages for business companies in most industries. Therefore, successful management requires that company's intangibles must be identified, measured and controlled [3]. That is why we encourage a growing interest in the new techniques of intellectual capital accounting as a method of measuring and reporting the range of human and knowledge-based factors that create sustained economic value [30]. The concept of intellectual capital, however, can also be placed in the context of a more general expansion of management knowledge, which is increasingly assuming the role of a "self-sustaining and self-reinforcing system" of integrated knowledge sources [30].

Underpinning the current interest is the claim that knowledge-based or intellectual assets, such as human capital, increasingly provide crucial sources of economic value. A discourse is under construction that expresses a sense of urgency about adopting these new approaches, and a sense of a compelling need to manage intellectual capital for sustained value creation [30]. At the same time, because this kind of organizational knowledge is tacit and intuitive, controlling it is highly problematic. The resources and assets involved have long been regarded as lying outside the reach of traditional accountancy in its role of measuring and reporting business performance. Indeed, attempts to devise new ways of making 'hidden value' visible have met with a growing awareness that the dynamics of value creation are not being adequately represented. Such an environment of rapidly changing ideas, coupled with ambiguity and unease about established ways of operating, is precisely the kind of scenario in which new forms of management knowledge expand [30]. And this is an opportunity for HCA to fulfill its aims of providing information on how human capital is creating value for the organization.

6.5 Reporting on Human Capital

Studies on the usefulness of intangibles for policy-making purposes have identified a number of critical changes in our societies (such as the globalization of business activities, the intensification of competition, and the unprecedented development of information technologies) as a result of which knowledge has become the fundamental production factor. Research has demonstrated that productivity gains are mainly driven by the use of knowledge and that intangibles are the key drivers of competitiveness [3, 29]. When an company's management is not aware of what its intangible assets are, it can miss business opportunities based on these intangible resources, because managers will be making key decisions without taking them into account among possible variables [3]. Thus, the move to a knowledge-based economy demands the emergence of an accurate measurement of the value of knowledge in a company: a broader asset framework. Managers must recognize that there are intellectual capital assets, such as highly skilled employees, that need to be identified and managed. There are human assets, not in the sense as normative accounting gives to the concept of 'asset' but management need information about top management quality, top management experience, ability to execute on strategy, leadership capabilities, problem solving ability, employees loyalty (behavioral and attitudinal), personnel reputation, workforce adaptability and employee engagement [29]. However, the revision of current financial accounting principles and standards would only be justified if it improves the usefulness of accounting information and leads to the overcoming of important problems facing companies and their stakeholders. Therefore, researchers have devoted effort to the identification of a number of significant damages resulting from the lack of publicly available relevant and reliable information on intangibles (e.g. [32]). Recent studies have shown that the lack of information on intangibles may increase uncertainty, and lead to the undervaluation of companies and the existence of greater errors in analysts' earnings forecasts [18, 31].

The Skandia Group, an international corporation that offers insurance and financial services, has experimented HCA since its vice-president Leif Edvinsson noticed that as much a company invested in their human capital, the lowest value was presented in the financial statements, because human capital/resources are treated as an expense in the Income Statement. Under the direction of Edvinsson, the Assurance and Finance Services Division pioneered the field of knowledge management by creating the first intellectual capital supplement to a corporate annual report in 1991 [2, 33]. Designed to reflect the value of intellectual capital within the organization, it measured the impact that human capital had on shareholder-owned structural capital, and gauged how intellectual assets had been leveraged over the preceding year. This experience has led to the Skandia Navigator concept. This concept was designed to provide a balanced picture of the financial and intellectual capital within an organization (since it focuses on four intellectual capital areas: customer, process, human, and research and development). In addition to providing a general overview of intellectual capital, the Skandia Navigator provides a management process with which to develop and predict the future value of this capital [2, 33].

The work at Skandia developed by Edvinsson [33] and of thinkers such as Sveiby [34] and the Roos brothers [35] has demonstrated the inadequacy of the traditional balance sheet to reflect the health of a company, but has also looked for alternatives. According to Mayo [5], the work developed by these authors is also important for those who work as professionals in HRM for several reasons. The reasons are: if taken seriously, it balances the focus of a company, between money and people, between short and long term; it provides an opportunity for HRM to be firmly linked to the bottom line and to the agenda of top management; it helps HRM to prioritize activities according to the level of value that will be contributed to the organization.

The concern for intellectual capital is different from the balanced scorecard, according to Mayo [5]. Kaplan and Norton [36] highlighted the inadequacy of financial measures alone as indicators of performance and success. They argued that financial measures should be balanced with those relating to customers, innovation and learning, and internal efficiency. But the balance may not reflect the true strategic drivers of value (that is, creating today's wealth but also generating the capability of tomorrow's wealth). It is the dynamics of growth, rather than mere measurement itself, that makes a difference [5].

Knowledge management itself emerged a couple of years earlier than intellectual capital, but reflects the same broad changes: the rise of the knowledge economy; the role of information; and the creation and leveraging of knowledge assets [37]. In order to enhance knowledge management, it is important that the underlying division of managerial labor is rejected in favor of an integrated pursuit of measurement, management and reporting. Accountants are not to be concerned with only the measurement and reporting aspects, leaving their managerial colleagues the task of management itself. All three aspects should be of interest to both parties. In this way, knowledge management and intellectual capital, as contemporary management themes, are best understood as being two sides of the same coin [37].

The demand for better information about human capital has been evident during the 1990s. This interest has been shown by many different parties (e.g. human resource departments, financial departments, company doctors, unions and, more recently, from top management, investors and politicians) [20]. The disclosure of information on organizational knowledge resources and related knowledge management activities in annual reports has become a much debated issue within the intellectual capital discourse. Boedker, Guthrie and Cuganesan [4, 36] contrasted and compared the case study of organization's internal intellectual capital management issues and practices with its external intellectual capital reporting practices. Their empirical analysis demonstrated inconsistency between those variables. It showed that strategically important information about the organization's management challenges, knowledge resources, knowledge management activities and intellectual capital indicators were not disclosed to external stakeholders in the organization's annual reports [4]. An increasing body of literature is documenting a high payoff from human capital investments. However, not all the empirical studies reveal the importance of human capital, namely the low ranking of human capital indicators. And the reasons appointed are a knowledge problem (people are not aware of the payoff of these investments, the uncertainty and the ownership problem associated with human capital issues, and the insurance that management has the capability to take action upon data) [39]. Moreover, even if there are human capital indicators that can be disclosed, there is the problem of reliance. Do indicators of human capital transform adequate information? Are they valid? And are the methods of measurement reliable? These issues of validity and reliability could be referred to as the uncertainty problem and they are important from the stakeholders' point of view [40]. Besides, the users of that information on human capital do not know if the measures disclosed actually matter in the management control processes of the company, that is, if management takes the necessary action on data [40].

6.6 Human Capital Accounting as a Challenge to Both Accounting and HRM Fields

Some attempts have been made to put HCA on the political agenda.

Roslender and Stevenson [13] describe what happened in the UK with the *Accounting for People* initiative. In January 2003, the government announced the formation of a task force on human capital management, charged with considering how it might be possible to "account for people". According to the authors, the *Accounting for People* initiative was seen by several people as the possibility of a real step forward in promoting the interests of employees, and therefore, human capital accounting/management. Despite its distinctly managerial discourse, the initiative acknowledged that having now recognized that "people are our greatest asset", companies should begin to consider how to report on their human capital management activities in the annual corporate report. By doing so, they were as well as contributing to increased transparency in financial reporting, since mandatory reporting on human capital management affords considerable support to those individuals and organizations who believe that people are now the most valuable asset available to management. However, and according to Roslender and Stevenson [13], after more than three and half years of consultation, debate and deliberation, no substantial evolution has been made because larger UK quoted companies continued to be charged with providing a minimal level of general information on their employees [13]. This output laid down all the expectations for evolution where HCA reporting is concerned.

Indeed, several authors were advocating that: managers were willing to know more about intangibles in order to be able to improve their information systems and control mechanisms; companies were disclosing increasing amounts of voluntary information on their intangibles; managers were not only concerned for what accounting standards currently require or would require in the future; and companies need to communicate with their stakeholders, be transparent and trustworthy [31]. However, when we compare these arguments to the experience reported by Roslender and Stevenson, we conclude that we should be worried about the real intentions of management. On the one hand, they are aware that human capital is the source of competitive advantage, that knowledge and innovation management are crucial for the company's success. On the other hand, they seem to be unwilling to develop human capital performance and measurement systems because ambiguity gives them flexibility in order to manage at their free will. That is why we claim that the accounting discipline plays a crucial role in developing HCA to the benefit of the whole society and all the stakeholders of the company. However, it is necessary that the discipline should be free of the powerful sectional interests mentioned by Roslender and Stevenson [13].

Most senior managers intuitively understand that human capital has the potential to be strategically important. There is little beyond anecdotal evidence, however, to demonstrate its impact on financial performance, much less the contribution of human capital. Becker, Huselid and Ulrich [23], experienced researchers in this field, showed a clear relationship between what is called the high performance human resources systems and various measures of company financial performance. However, for management and reporting purposes, companies need to implement a performance measurement system, and develop measures that, on the one hand, monitors employee motivation because it is the basis for developing skills and increasing effort, the end point of which is the company's increase of profitability; and, on the other hand, to control for compensation, culture, management style and also job design because they give support to the employee motivation/satisfaction. According to Atkinson, Waterhouse and Wells [1], this performance measurement system applied to human capital must do four things: (1) help the company evaluate whether it is receiving the expected contributions from employees; (2) help the company evaluate whether it is giving to human capital what it needs to contribute to the increasing of profitability of the company (by monitoring this element, the company can identify problems quickly and correct them before they significantly affect the organizational goals); (3) guide the design and implementation of processes that contribute to the company's compensation, culture, management style, and also job design; and (4) help the company in evaluating its planning and the contracts, both implicit and explicit, that it has negotiated with human capital by helping evaluate the effect of compensation, culture, management style and also job design on human capital satisfaction/motivation, and in turn the financial goals of the organization.

For companies to improve performance, they must develop a comprehensive system of measures that monitor and evaluate the ability of its processes to achieve employee motivation. By focusing on results, rather than on their causes, the company resigns itself to being reactive rather than proactive in meeting the need for organizational change. Therefore, the performance measurement system is a vital management system that includes both financial and nonfinancial measures of performance [1]. The performance measurement system supports the development of organizational knowledge and leads to more systematic, effective decision making. While the process is not easy, it is necessary for managing and improving a company's performance.

The HCA framework allows management access to better information to help determine where it should focus the human capital investments in order to produce the greatest business benefits. Management base investment decisions on three criteria: the relative effectiveness of each human capital process, the presence or absence of evidence that a given process was strongly associated with financial performance, and the strategic or cultural importance of each process based on its linkage to particular key performance drivers of human capital capabilities [25].

Delivering performance is more than just having individuals with necessary capability [5]. Company performance is the result of the mixture of individual capability, individual motivation, leadership, the organizational climate and work effectiveness.

Individual capability is the set of knowledge, skills, experience, and the sense of network, the ability to achieve results, and the potential for growth of the individual. Its motivation, aspirations, ambitions and initiative, the work motivation, productivity, along with the capabilities, potentiates performance. According to Mayo [5], studies show repeatedly that the most important motivator is the nature of the work itself and its appeal and interest to the person. Matching people to roles that bring interest and enjoyment is thus a key process and management's responsibility. At the same time, leadership is also important (the clarity of vision of top management, their ability to communicate it, and the consistency of their behavior). Another factor that promotes performance is organizational climate, that is, the culture of the organization, mainly in its freedom to innovate, openness, flexibility, and respect for the individual. And finally, work effectiveness, that is, supportiveness, mutual respect, and sharing in common goals and values. The organizational climate can be measured by questioning employees, namely through surveys [5].

According to Johanson [39], measuring and reporting intangibles are one thing, but to ensure that understanding and action are based on the information from intangibles indicators, a number of supporting processes need to be performed. These processes are: the recognition and measurement routines (human capital surveys, market capital surveys, accounting); reporting routines (continuous internal reports, informal information to analysts); evaluation routines (evaluation of single indicators by each manager, statistical analysis); attention routines (meeting); motivation routines (benchmarking, dialogues, salary bonus); commitment routines (ownership of indicators, action contract); and follow-up routines (statistical analysis). Measuring without doing anything is worthless or even dangerous [39].

Although both researchers and managers have been talking extensively about intellectual capital for almost two decades, it appears that people still do not fully understand what it encompasses [3]. Although it is fully appreciated that intellectual capital or human capital can provide substantial competitive advantage, managers do not fully understand what it is and how it works. This may be particularly so in the context of how investments in human capital have impact on the operation of a business [3]. Therefore, due to the concern and awareness of the importance of human capital as a strategic organizational resource as a key asset in the innovation process, it is time for management to take action. With the assistance of accounting professionals, management implements performance measurement systems capable of providing information to manage the human capital, to report to the stakeholders, and to respond to questions related to company's social responsibility and sustainability.

However, many companies are likely to remain reluctant to publish such information, fearing that action of this sort may result in exposing a company in two different ways. In the first place, for many firms nowadays these assets are part of their core business; therefore, disclosing too much about them might reveal the company's competitive advantage. They may prefer to withhold this information from competitors until they are required to do so [40]. Although this argument makes sense, it seems it is used to camouflage the absence of an HCA system, capable of providing information of this core asset, and how it is considered in the company's management process, namely how it is related with company's performance. Besides, since financial accounting has not changed where human assets is concerned, and requirements to disclose information on intellectual capital are still voluntary, information is provided only at management's will. Therefore, arguing that the reporting model does not require it, it is the same as stating that companies would not disclose this information at their own will, and they boycott any attempt to change this – see what happened with the *Accounting for People* initiative reported by Roslender and Stevenson [13]. Secondly, intellectual capital reporting might not only expose the bases of competitive advantage but may provide clues as to a company's weaknesses. Revealing information of this sort might provide problems for managers, not only because their competitors can act or perform better, but also with internal stakeholders, who now realize that, in fact, the business is not performing quite as well as they thought it was [3]. This is the problem that had been already identified early in the 1960s: CEOs stating that "our employees are our most important!– and most valuable – asset", providing no other information about it in the rest of the annual report [7]. Moreover, it could put in evidence that managers' action is not in accordance with the managers' discourse.

Fortunately for managers, organizations that choose to make information available on human capital can decide what is disclosed and how it is communicated. The history of social and environmental accounting developments has demonstrated how such unregulated accounting spaces can be populated to the benefit of those whose motivations might not withstand rigorous scrutiny [3].

Some authors speak of the competition between accountancy and HRM professions for organizational dominance, namely that this phenomenon could explain the failure of the *Accounting for People* initiative [13].

Armstrong [41] documents how the accounting and personnel functions, among others, have been involved in a process of interprofessional competition for organizational dominance, with the former traditionally enjoying greater success. The passage of time, and the emergence of the (strategic) HRM profession, has done little to change this situation of no recognition of human capital assets in the company's reporting, despite the acknowledgment that "our people are our greatest asset" [13]. In fact, the progress in accounting for people has been rather limited during the last 50 years, with the exception of the Scandinavian context. Indeed, for the greater part, accounting for people has been principally an academic preoccupation, attracting very little interest in the accountancy profession. However, there are many indications of the awe in which the accountancy profession is held by senior management, and there is a sense that the accountancy profession alone is deemed capable of providing the necessary measurement metrics and that it also possesses the authority to provide the legally sanctioned external report of organizational performance [13]. However, we claim that it is not possible to evolve HCA and reporting without the contribution of the human capital management field. When questioning why the *Accounting for People* initiative disregarded recent developments in intellectual capital that could be of positive benefit to progressing HCA, Roslender and Stevenson [13] answered that those developments threaten to promote the interests of labor (people) within the prevailing social arrangements because it poses a problem to capital (senior management), and the accountancy profession does nothing because management is its most important client.

In an early study by Rhode and Lawler in 1973 [26] that focused on the usefulness of HRCA, it was found that managers opposed HRCA because it was perceived to limit their freedom of action. Organizational learning processes were initiated though seriously hampered by management's ambivalent support of HRCA. As Rhode and Lawler reported [26], the added clarity provided by HRCA implies that a manager's freedom of action, and therefore power, is circumscribed. HRCA can also be employed to assess the efficiency of the managers themselves. These may be reasons for its lukewarm support [20].

Managers have to learn about HRCA models (e.g. How can the cost of absenteeism from work be calculated? What models can be used to estimate the financial impact of a training program?). Thus, two types of knowledge have to be improved [20]: (1) knowledge of human resource costs and values or the normal outcome of different human resource measures within the organization, and therefore appropriate information systems are needed; (2) knowledge of models of how to calculate costs, incomes and values. Lynn cited by Fincham and Roslender [30] claims that management accounting has a 'natural affinity' with intellectual capital accounting and that the movement to manage intellectual assets is fundamentally a management accounting issue. In like fashion, Booth cited by Fincham and Roslender [30] seeks to convey the message that the measurement of intellectual capital is actually unproblematic and falls within conventional intellectual capital accounting as management fashion accounting skills. He suggests that a range of off-the-shelf methods are ready to be simply plugged in. The real difficulties lie in the managerial processes of understanding intellectual capital and using it to improve organizational performance – tasks at which most managers probably think they excel – but the "modelling tools are readily available and computational requirements rarely a hurdle" [30].

6.7 Conclusions

The preceding pages have documented how human capital evolved in two areas, measurement and reporting, and how human capital is important not only in the context of strategic HRM, but also as a factor in knowledge/innovation management, and the intellectual capital literature. Although facing several designations, HCA has not achieved yet the status that it aims for, that it deserves, and that it is crucial. While empirical research studies show that it is vital for senior managements to view intellectual capital in a strategic way if they wish to maximize the benefits that these assets can bring to their organizations, it is also vital to manage them effectively. However, we can manage only what we can measure. Therefore, from a specifically accounting perspective, the task has been to identify appropriate ways of measuring and reporting the success with which stocks of intellectual capital have been grown over time. In this regard it seems unlikely that accounting, as it has traditionally been understood, is capable of meeting these new challenges [3].

The conclusion is that performance measurement systems are still based primarily on financial performance measures, which lack the focus and robustness needed for internal management and control [1]. Financial performance measures are derived from accounting systems that generate and communicate financial information to support the contractual relationships and the capital markets that result from separating owners and managers in the modern corporation [1]. These accounting systems were designed with the priority of consistency and hardness – attributes needed to instantly compare firms and evaluate a company's behavior over time. These systems were not designed to communicate decision-relevant information to people inside the organization. Some complaints raised with conventional financial information are that it ignores important issues like human capital or customer satisfaction, cannot predict because it is based on historical cost, and provides little or no basis to judge the effectiveness of processes like personnel relations systems [1].

A company's success is created by monitoring and managing its performance on the compensation, management style, organizational culture, and job design, since success in achieving performance on increasing profitability follows from the former variables. Therefore, performance measures of those variables are the way to improve the company's financial performance and should be the focus of the company's measurement [1].

A HCA framework can help organizations diagnose their strengths and weaknesses in key human capital practices, to set investment priorities and track performance, and to establish an empirical link between human capital investments, business practices, and overall business performance [25]. Most managers would agree, however, that mature people processes do not immediately impact financial performance; they first improve human capital capabilities like employee engagement and workforce performance, which in turn improve key performance drivers like customer satisfaction and innovation [25].

A HCA framework provides the human resources function and the business leaders with a common language and vocabulary concerning the value of human capital investments and the contribution to the bottom line [25]. The only way for the human resource function to be successful is to learn to think and act like a business person [25].

Companies seek to share knowledge and provide information to their various stakeholders; in order to incorporate information on intellectual capital, they need redesign their information and reporting systems [31].

Any lack of visibility of intellectual capital within accounts, especially on human capital, results in information asymmetries favoring those who have privileged access to that information, because they work within the organization. It is unknown data for those who are not involved in the management of the company. As a result, a management team might report unusually positive performances, while competitors report losses [3]. Furthermore, when a company values and reports its intangible assets, its capacity for raising capital increases [3].

Accounting encompasses two complementary activities: measurement and reporting. Monetary valuations are the archetypal measurement metric associated with accounting, while the balance sheet has traditionally provided the vehicle for reporting the aggregated valuations of tangible together with some intangible assets. Denied the former in the case of intellectual capital, the latter was of little use for such purposes. The new approaches might usefully be categorized into three generic types:

alternative hard number metrics; scoreboards populated by sets of softer indicators; and narrative accounts of intellectual capital growth in which indicators performed a largely supplementary role [3].

In the case of accounting for intellectual capital the most telling issues again refer to human capital. Visualizing the growth of specific employee attributes in the form of nonfinancial metrics to be incorporated into some form of scoreboard reporting framework remains problematic on the grounds that, from a critical accounting perspective, accounting numbers applied to employees have invariably worked to their disadvantage [3]. However, if HCA wants to be part of the management process, and in the disclosure process to all the company's stakeholders, it is necessary that those numbers that allow management and stakeholders to assess the company's practices and its financial performance.

In summary, while 'employee matters', together with those relating to society, the community and the environment, continue to be invoked by many people, it is difficult to avoid concluding that, after a brief interlude when encouraging accounting for people was on the political agenda, it is once again a very minor consideration [13]. Accounting for people continues to be resultantly represented as a largely technical problem, a solution to which might eventually be found within the prevailing conceptual framework of financial accounting and reporting [13]. The great attraction of this situation is that as long as it persists, it provides senior management with a rationale for continuing to account for people as costs to the company rather than as assets. And therefore, HCA is nothing more than a beautiful management and accounting fairy tale.

Although the result of the *Accounting for People* initiative reported by Roslender and Stevenson [13] comes as no surprise, we share the authors' view that the existence of renewed interest in accounting for people through a focus on intellectual capital ultimately provides critical accounting and HCA scholars with grounds for a degree of optimism. Research in the intellectual capital field is being pursued against a background of growing interest in developing a strong critical orientation across the whole spectrum of management studies. Increasing numbers of scholars may therefore be more likely to be receptive to arguments that promote the interests of those who are managed, designated 'human capital' in the intellectual capital field [13].

Bibliography

[1] A. A. Atkinson, J. H. Waterhouse, and R. B. Wells, "A Stakeholder Approach to Strategic Performance Measurement," *Sloan Manage. Rev.* vol. 38, no. 3, pp. 25–38, 1997.

[2] E. G. Flamholtz, M. L. Bullen, and W. Hua, "Human resource accounting: a historical perspective and future implications," *Manag. Decis.* vol. 40, no. 10, pp. 947–954, Jan. 2002.

[3] L. Alcaniz, F. Gomez-Bezares, and R. Roslender, "Theoretical perspectives on intellectual capital: A backward look and a proposal for going forward," *Account. Forum*, vol. 35, no. 2, pp. 104–117, Jun. 2011.

[4] C. Boedker, J. Guthrie, and S. Cuganesan, "The Strategic Significance of Human Capital Information in Annual Reporting," *J. Hum. Resour. Costing Account.* vol. 8, no. 2, pp. 7–22, 2004.
[5] A. Mayo, "The role of employee development in the growth of intellectual capital," *Pers. Rev.* vol. 29, no. 4, pp. 521–533, Jan. 2000.
[6] B. Becker and M. Huselid, "Measuring HR?," *HR Mag.* vol. 48, no. 12, pp. 56–61, 2003.
[7] R. L. Brummett, E. G. Flamholtz, and W. C. Pyle, "Human Resource Management – A Challenge for Accountants," *Account. Rev.* vol. XLIII, no. 2, pp. 217–224, 1968.
[8] J. S. Hekimian and C. Jones, "Put people on your balance sheet," *Harv. Bus. Rev.* vol. January-Fe, pp. 105–113, 1967.
[9] R. H. Hermanson, "Accounting for Human Assets," Michigan State University, 1964.
[10] E. G. Flamholtz, *Human Resource Accounting: Advances in Concepts, Methods, and Applications*, 3rd edn. Kluwer Academic Publishers, 1999, p. 390.
[11] B. Lev and A. Schwartz, "On the use of economic concept of human capital in financial statements," *Account. Rev.* vol. january, pp. 103–112, 1971.
[12] A. Friedman and B. Lev, "A Surrogate Measure for the Firm's Investment in Human Resources," *J. Account. Res.* vol. Autumn, pp. 235–250, 1974.
[13] R. Roslender and J. Stevenson, "Accounting for People: A real step forward or more a case of wishing and hoping?," *Crit. Perspect. Account.* vol. 20, no. 7, pp. 855–869, Oct. 2009.
[14] B. Marr, G. Schiuma, and A. Neely, "The dynamics of value creation: mapping your intellectual performance drivers," *J. Intellect. Cap.* vol. 5, no. 2, pp. 312–325, 2004.
[15] N. Elias, "The effects of Human Asset Statements on The Investment Decisions: an Experiment," *JEm pirical Res. Account. Sel. Stud. Suppl. to J. Account. Res.* vol. 10, pp. 215–233, 1972.
[16] J. A. Hendricks, "The impact of Human Resource Accounting Information on Stock Investment Decisions," *Account. Rev.* vol. 51, no. 2, pp. 292–305, 1976.
[17] E. Schawn, "The effects of human resource accounting data on financial decisions: an empirical test," *Accounting, Organ. Soc.* vol. 1, no. 2–3, pp. 219–237, 1976.
[18] B. Hansson, "Personnel investments and abnormal returns: knowledge-based firms and human resource accounting," *J. Hum. Resour. Costing Account.* vol. 2, no. 2, pp. 9–29, 1997.
[19] I. Moreno-Campos, "La información Contable sobre intangibles: El caso del Capital Humano. Evidencias empíricas," University of Seville, Spain, 2000.
[20] U. Johanson, "Why the concept of human resource costing and accounting does not work A lesson from seven Swedish cases," *Pers. Rev.* vol. 28, no. 1/2, pp. 91–107, 1999.
[21] B. Gerhart, P. M. Wright, and G. C. Mcmahan, "Measurement Error in Research on the Human Resources and Firm Performance Relationship: Further Evidence and Analysis," *Pers. Psychol.* vol. 53, no. 4, pp. 855–872, Dec. 2000.
[22] B. E. Becker and M. A. Huselid, "High performance work systems and firm performance: A synthesis of research and managerial implications," in: *Research in personnel and human resource management*, G. Ferris, ed., vol. 16, 1998, pp. 53–101.
[23] B. E. Becker, M. Huselid, and D. Ulrich, *The HR Scorecard – Linking People, Strategy, and Performance.* Boston, MA: Harvard Business School Press., 2001.
[24] E. G. Flamholtz, D. G. Searfoss, and R. Coff, "Developing Human Resource Accounting as a Human Resource Decision Support System," *Account. Horizons* vol. 2, no. 3, pp. 1–9, 1988.
[25] S. Cantrell, J. M. Benton, T. Laudal, and R. J. Thomas, "Measuring the value of human capital investments: the SAP case," *Strateg. Leadersh.* vol. 34, no. 2, pp. 43–52, Jan. 2006.
[26] J.-E. Gröjer and U. Johanson, "Current development in human resource costing and accounting: Reality present, researchers absent?," *Accounting, Audit. Account. J.*, vol. 11, no. 4, pp. 495–506, Jan. 1998.
[27] B. Quinn, P. Anderson, and S. Finkelstein, "Managing professional intellect: Making the most of the best.," *Harvard Bus. Rev. Mar/Apr96* vol. 74, no. 2, pp. 71–80, 1996.

[28] L. Edvinsson and M. S. Malone, *Intellectual Capital*. New York: Harper Business, 1997.
[29] J. J. Ballow, R. Burgman, and M. J. Molnar, “Managing for shareholder value?: intangibles, future value and investment decisions,” *J. Bus. Strategy* vol. 25, no. 3, pp. 26–34, 2004.
[30] R. Fincham and R. Roslender, “Intellectual capital accounting as management fashion: a review and critique,” *Eur. Account. Rev.* vol. 12, no. 4, pp. 781–795, Dec. 2003.
[31] M. García-Ayuso, “Intangibles: Lessons from the past and a look into the future,” *J. Intellect. Cap.* vol. 4, no. 4, pp. 597–604, Jan. 2003.
[32] B. Lev and P. Zarowin, “The boundaries of financial reporting and how to extend them,” *J. Account. Res.* vol. 37, no. 2, pp. 353–385, 1999.
[33] L. Edvinsson, “Developing Intellectual Capital at Skandia,” *Long Range Plann.* vol. 30, no. 3, pp. 366–373, 1997.
[34] K.-E. Sveiby, *The New Organizational Wealth: Managing and Measuring Knowledge-Based Assets*. San Francisco: Berret-Koehler Publishers, 1997.
[35] G. Roos and J. Roos, “Measuring your Company’s Intellectual Performance,” *Long Range Plann.* vol. 30, no. 3, pp. 413–426, 1997.
[36] R. Kaplan and D. Norton, *The Balanced Scorecard*. Boston, MA: Harvard Business School Press, 1996.
[37] R. Fincham and R. Roslender, “The Management of intellectual capital and its implications for business reporting,” Edinburgh, 2003.
[38] C. Boedker, J. Guthrie, and S. Cuganesan, “An integrated framework for visualising intellectual capital,” *J. Hum. Resour. Costing Account.* vol. 6, no. 4, pp. 510–527, 2005.
[39] U. Johanson, “Why are capital market actors ambivalent to information about certain indicators on intellectual capital?,” *Accounting, Audit. Account. J.* vol. 16, no. 1, pp. 31–38, Jan. 2003.
[40] J. Holland and U. Johanson, “Value-relevant information on corporate intangibles – creation, use, and barriers in capital markets – ‘between a rock and a hard place,’” *J. Intellect. Cap.* vol. 4, no. 4, pp. 465–486, Jan. 2003.
[41] P. Armstrong, “Changing management control strategies: The role of competition between accountancy and other organisational professions,” *Accounting, Organ. Soc.* vol. 10, no. 2, pp. 129–148, Jan. 1985.

Jorge da Silva Correia Neto, Jairo Simião Dornelas*, and
Andrea Gomes Santos

7 Beyond the 3C Model in Collaboration Platforms: A Case Study

Abstract: Whereas the information society emerged supported by the Web, the collaboration society is emerging with the Web's online applications, platforms and media, which formed what is known as Web 2.0 and empowered interactivity and participation. In the 1990s, the 3C collaboration model introduced the collaboration construct as formed by the dimensions of communication, coordination and cooperation. However, using this model to guide the development of modern artifacts of collaboration, such as social network sites and interactive collaboration platforms, raises the following question: Is it possible to identify something beyond the 3C model through the analysis of features provided by collaborative platforms? This study presents a preliminary analysis that posits a fourth dimension to the collaboration construct: the interactivity dimension.

7.1 Introduction

Whereas the World Wide Web, or simply the Web, created what is known today as the information society, the Web 2.0, with its online applications, platforms and media, is leading to the emergence of the collaboration society, as indicated by Tapscott and Williams [1]. Since the late 1990s, a new product family (wikis, podcasts, blogs and social networking sites, among others) began to settle what is known today as Web 2.0 or the social web, which has as its main differential a new kind of interactive and participatory environment [2].

In addition to communication and interactivity, another key concept in Web 2.0, as already emphasized, is collaboration. Peng and Woodlock [3, p. 202] point out that a person must be constantly collaborating to create, share and distribute value and capabilities with others, because "collaborating social actors integrate knowledge and expertise from different sources, trigger new ideas that challenge the current understanding and provoke the emergence of new solutions." Thus, in this communication universe powered by Web 2.0, as stated Vreede, Briggs and Massey [4, p. 121], the collaboration becomes "a critical phenomenon in organizational life." Therefore, many organizations have begun to appropriate the dynamics of social networks for

Jorge da Silva Correia Neto: jorgecorreianeto@gmail.com
***Corresponding Author: Jairo Simião Dornelas:** Federal University of Pernambuco, Administrative Science Department. Email: jairo@ufpe.br
Andrea Gomes Santos: andrea.g.santos@gmail.com

increased synergy between their employees [5] and between business partners [6], i.e. with intra- and inter-organizational developments.

When collaboration involves a large number of people to obtain services, ideas, contributions or content, with extremely low cost or voluntary actions, building something that is beneficial to the community as a whole, this practice is known as crowdsourcing, as posed by Howe [7] and Doan, Ramakrishnan and Halevy [8].

New challenges arise for the implementation and use of these tools, which go beyond expanding the communication with the market. In fact, these challenges are also used for the co-creation and co-production of goods and services. As pointed out by Ramaswamy and Gouillart [9, p. 4], the consumer experience is "central to value creation, innovation, strategy and leadership," because success lies on using the experiences of this engaged community to generate ideas that improve the nature these interactions.

In order to develop their collaborative platforms, organizations can use one of the most recognized models in the academic world, the 3C collaboration model [10]. Originally proposed in the early 1990s by Ellis, Gibbs and Rein [11], this model used a slightly different terminology. Nowadays, the 3C model [12] involves communication, coordination and collaboration as the three dimensions of the systems that support group work.

Researchers conceptualize collaboration as being formed by the dimensions of communication, coordination and cooperation [10]. Thinking about how a model originated in the 1990s is still being used to guide the efforts of the development of collaborative platforms such as these social and interactive platforms raises the following question: When analyzing features provided by current collaborative platforms, is it possible to identify dimensions beyond the ones present in the 3C model?

To investigate this question, the present chapter is structured in three sections that follow this first and introductory section. The second section presents a literature review; the third section presents the methodological procedures. Finally, the fourth section presents the results section and final considerations.

7.2 Literature Review

This section will explore the concepts within the 3C model of collaboration and interactivity, aiming to discuss the main components of the dimensions present in the current collaboration platforms.

7.2.1 3C Collaboration Model

Generically, collaboration involves two or more people sharing complex information in pursuit of a common goal or purpose [13]. Corroborating with this view, Briggs et al. [14] argue that collaborative efforts are conducted jointly and target a mutual goal, regardless of individual positions on tasks to be performed or goals to be individually pursued. Moreover, collaboration "also connotes a more durable and pervasive relationship than the rudimentary level afforded by simple interaction" as Harley [15, p. 64] explains.

In the 3C collaboration model, proposed by Fuks et al. [12], collaboration is seen as a combination of cooperation, communication and coordination. The present study envisages the possibility of adding to this model the interactivity dimension, which is seen as crucial to the understanding of collaboration in the context of Web 2.0.

In the 3C collaboration model, "communication is related to the exchange of messages and information between people; coordination is related to the management of people, their activities and resources; and cooperation, which is the production being held in a place shared" [12, p. 637]. Figure 7.1 shows how these dimensions interact with each other to support and empower collaboration to create a shared workspace.

Fuks et al. [12], say that the 3C model should not only be used to classify collaborative systems, but especially to implement groupware (collaborative software), providing the features mapped in its key aspects, or dimensions. According Fuks et al. [16], collaborative systems should consider the following dimensions of collaboration:

- Coordination: There are three kinds of coordination. Coordination of people is related to communication and its context; coordination of resources is linked to the shared environment where interactions occur; and coordination of the work involves the management of interdependencies between tasks performed to achieve a common goal;
- Cooperation: the members of a group cooperate in producing, manipulating and organizing information and, also, building and refining the objects cooperatively;
- Communication: the media to be transmitted must be defined (e.g. text, speech, images, etc). The transmission mode (synchronous/asynchronous); restriction policies (text size or time of the videos), meta-information (such as title, date, priority, category, etc.) and the structure of conversation (linear, hierarchical or network) must also be defined.

Besides these dimensions in this collaborative context, the perception, a new variant, stands out, and helps to understand the activities of the other group members and how they are developing in the shared space, serving as a positive or negative indicator of the individual performance enhancement.

Finally, as shown in Figure 7.1, the record of group activities are completed, cataloged, categorized and structured around the cooperation objects. Ideas, facts, issues, views, conversations, discussions, decisions and so on, could be recovered, providing historical context and collaboration about the activity that took place in that shared environment. However, to the literature presented here, the perception of the activities carried out by other actors is just part of the interactive collaboration process; other

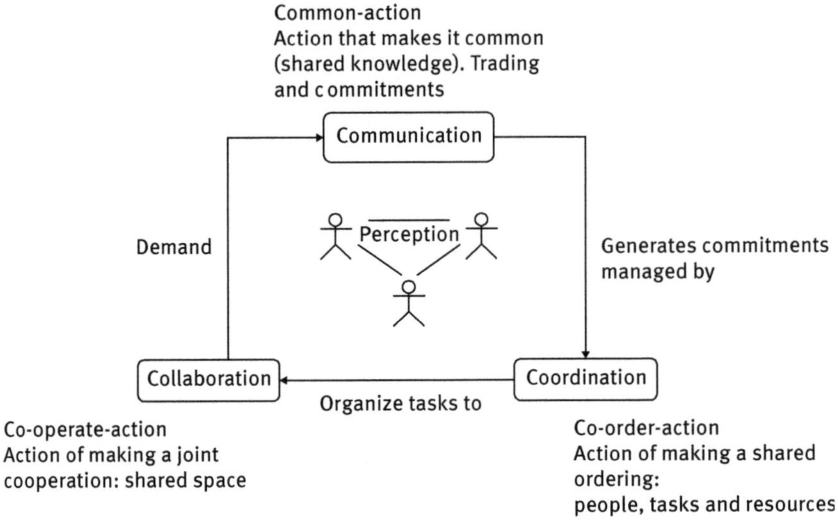

Fig. 7.1. Interacting in a shared space work. Adapted from Fuks et al. [12].

social factors also appear to contribute to the exchanges that occurred in this shared space, as will be shown in the next section.

7.2.2 Interactivity

The term interactivity emerged in the mid-nineteenth century and came from the idea of "two people or things affecting or causing any effect on the other; enabling two-way flow of information between a computer and a user, responding to a stimulus from this user" [17]. It can also be seen as a "resource, environment or communication process that allows the receiver to actively interact with the user" [18, p. 484].

Regardless of etymological perspective, Lemos [19] considers interactivity as more specifically digital interactivity, as a specific case of interaction, a kind of relationship between technological and social, a dialogue between man and machine, mediated by graphical interfaces and real time. Kiousis [20, p. 355] also points out that interaction is composed "by the media and psychological factors, which vary in terms of communication technology, communication context and perceptions of the people involved."

Aiming to provide an explanatory overview of this concept, as displayed in Table 7.1, McMillan [21] defines interactivity based on its resources, processes and perceptions on the one hand; and with respect to the actors, among which the interaction takes place, on the other hand. These interactions can be human–human, human–computer and human–content.

As stated by Ramaswamy and Gouillart [9], models of collaboration must provide components to strengthen interpersonal relationships, for example, highlighting the

Table 7.1. The concept of interactivity in terms of resources, processes and perceptions. Based on McMillan [21].

	ELEMENTS	INTERACTIONS Human–Human	Human–Computer	Human Content
INTERACTIVITY	**Resources**	Chat e-mail	Navigation menus Search tools	Tools to facilitate personalized content Forms
	Process	Joining a chat Send/receive e-mail	Navigating a website Using a search tool	Creating a customized home page Finding news in multiple media
	Perceptions	Finding that the chat and e-mail facilitate communication Can be based on personal interest or involvement	Finding a site that is easy to control and engage with Can be based on experience with technology as well as interest and involvement	Finding a site that is easy to control and engage with Can be based on experience with technology as well as interest and involvement

contributions made by individuals in their cooperative activities. Therefore, beyond the basic dimensions – coordination, cooperation and communication – collaboration models should increase interaction between subjects and between subjects and objects when pursuing the desired group results, suggest these quoted authors.

Accordingly, Silva [22, p. 10] points out that the interactivity emerges as a result of this conversational computer technology, but it is also a result of a marketing dimension that seek to broaden the dialogue between producer and customer. These relations are a product of a social dimension that seeks autonomy and no longer passively accepts references that determine meanings, such as church, ideology, and so on. In this new socio-historical context, interactivity provided by computation creates a new communication mode in which the message is modifiable; the sender constructs a network (and not a single path) of territories to explore and the receiver handles the message as co-author, reinforcing the sense of "participation, intervention, bidirectionality and multiplicity of connections" [22, p. 13].

Silva [22] also indicates that there are three elements of interactivity. The first plea, the binomial interest-intervention, increases the presence of the public in the communication process and gives them the ability to become managers of media. It also changes the nature of the receiver, because receiver and transmitter change their statuses when the message content is presented and manipulated by both actors. The sender assumes the receptor's participation-intervention because participating is interfering in the message.

The second plea, the binomial bidirectionality-hybridization, is the critical key to the functionalist view of classical communication theory, in which the relationship between sender and receiver is unidirectional. This new scenario empowers a joint and participative production. Thus, there is a recursion between sender and receiver. Communication is a joint production of the sender and the receiver. Both poles encode and decode.

The third plea, the binomial potentiality-interchangeability, allows the creation of multiple articulatory network connections. It does not propose a closed message; in contrast, it provides information on network connections allowing people freedom of association and meanings to the receiver and produces wide freedoms and several possible (potential) narrative combinations (interchangeability).

7.2.3 Related Work

Assuming learning management systems (LMS) as a type of groupware system, it is possible to abstract the learning experience reported in Fuks et al. [24] with the LMS AulaNet, developed based on the concepts of the 3C model. In this context, the authors exemplify each dimension of the 3C model from the discussion of environmental features, discussing how the 3C collaboration model guided the definition of the software requirements.

In another study, conducted by Gerosa et al. [25], the development of groupware is analyzed with the lens of componentization. Based on the aforementioned 3C model, researchers analyzed the "sharing services such as chat, forum, calendar, file management, whiteboard, and so on; and the services themselves that share elements such as list of participants, session control, timing objects, permission control, and so on", reinforcing the use of this model for the development of collaborative platforms that contain the same functionality as the platforms of co-creation, which are the targets of this study.

Ferro and Heemann [26] reflect on the use of the 3C model for the design of collaborative services franchise networks, pointing to the relevance of this model for the design of more collaborative processes between the various stakeholders involved.

In the same direction of the work summarized here, this article aims to analyze the functionality of a system from the perspective of the 3C model. However, beyond the three dimensions of the model, we intend to investigate if on a collaborative platform is it possible to identify the presence of other elements, not represented in the 3C model, such as interactivity, as discussed by Silva [22] and raised here as an emerging conjecture.

To do this, first it was necessary to build a list of the features of the automaker Local Motors' co-creation platform. According to Anderson [27], Local Motors has been implementing, since 2007, one of the most successful experiments of co-creation in the world, especially in the automotive industry. In order to enlist these functionalities, the platform was directly analyzed to select and describe these features, similar to the work of Magalhães et al. [28].

7.3 Methodological Procedures

This research used an exploratory-descriptive qualitative approach in order to identify and analyze the functionalities available on Local Motors' co-creation platform, which was the studied case. Subsequently, the main features identified were analyzed by a team of experts who considered the intention of each functionality in the context of collaboration between users. It must be emphasized that this co-creation platform was designed by the company for its users to develop prototypes and designs of new vehicles. In this sense, the functionalities were captured from the platform in January of 2014.

Exploratory studies aim to provide greater familiarity with a problem, to make it more explicit, especially when dealing with an under-investigated topic or a topic that has not been previously addressed [29]. The case study is an empirical inquiry that investigates a contemporary phenomenon that does not have boundaries or context clearly evident and which requires multiple sources of evidence [30].

To sort the functionality of this collaborative context, a team of experts was formed, composed of: a teacher with 20 years of experience, with a Doctor in Business Administration and research focusing on decision making; an expert on groupware and collaborative technologies and leader of a research group in information systems; a teacher with over 20 years experience, with a PhD in biomedical engineering and leader of a research group in collaborative healthcare technologies; two PhD students in Business Administration with focus on information technology and teachers with four years of experience in research on open innovation collaboration and enterprise social networking platforms; and two master's candidates in Computer Science with a focus on software engineering and experience in design and development of a collaborative social network to support health.

The process of selection and classification of features occurred in four steps:

- First stage: a pair formed by the two master's candidates using a public profile on the platform selects the features available to the community as well as a description of the actions of these features;
- Second stage: the same pair, working independently, does the classification of features extracted by observing their actions and classifying them according to the settings of the 3C model and interactivity, described in Section 7.2 of this paper;
- Third stage: these masters and the doctoral candidates analyze the classified features and discuss it, trying to achieve a consensus, and adding more experience to the classification process in order to create a single list;
- Fourth stage: the constructed list goes through an evaluation process with the two senior researchers in order to discuss the final ranking of the features. Figure 7.2 illustrates the process of this method as applied to this research.

The research method applied in this work resembles, in some points, the Delphi method of qualitative research, in which a group of experts meet to discuss and enter

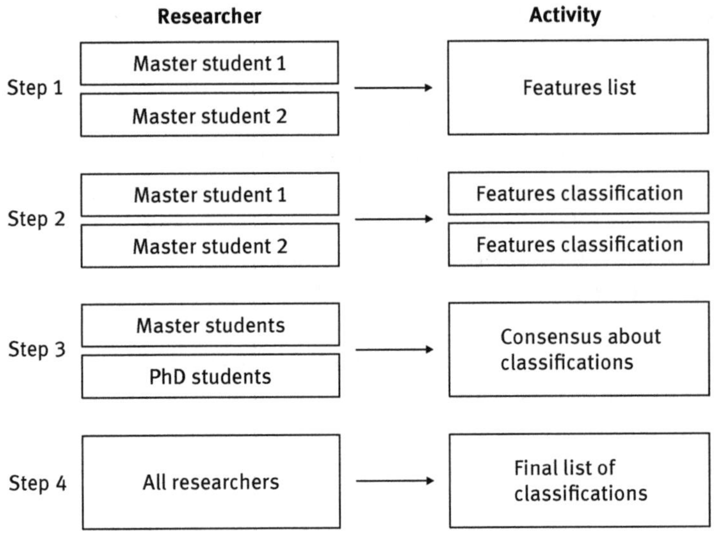

Fig. 7.2. Selection and classification of features procedure.

into a consensus on a particular issue [31]. Furthermore, the method also has some similarities with the formation and organization of a team of researchers to conduct a systematic literature review. In this process, the team is divided into two groups working independently in the extraction and classification of information. Subsequently, the group meets to discuss what has been accomplished and to try to find a consensus on divergent issues [23].

7.4 Results

In this section, the results of the research and the full discussion of the issues observed in the analysis of the collected information are presented.

7.4.1 Description of the Platform

Before discussing each of the platform's identified features, it is necessary to define some platform modules: *Ideas* are the basic features for each innovative project that may be created; *Designs* (or sketches) hold art drawings, renderings and models for the community; *Projects* are collaborative spaces where designers, engineers and enthusiasts come together to design and build solutions; and *Challenges* are competitions in which community members presents solutions to a given problem, with a chance to win a reward. Figure 7.3 shows the use case built to illustrate the main features available in the co-creation of the Local Motors platform.

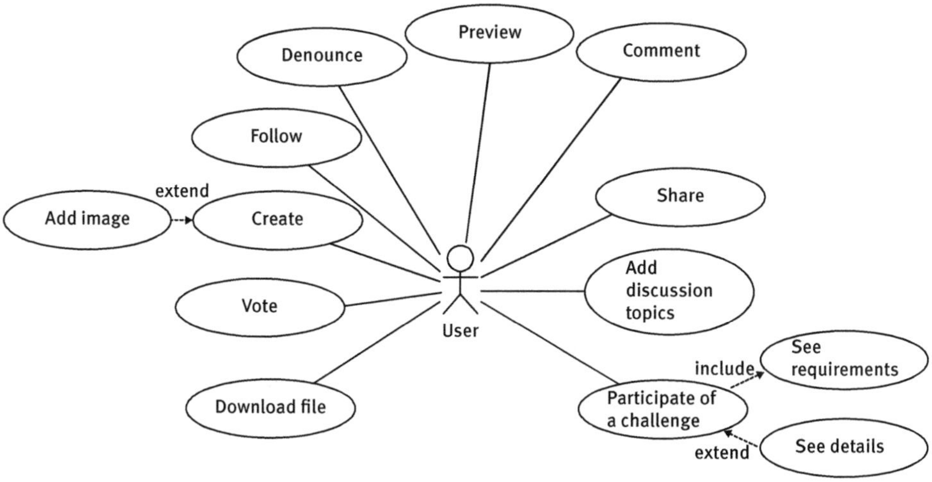

Fig. 7.3. Functions of Local Motors' co-creation platform.

Based on the use case diagram, Table 7.2 shows the functionalities available on the platform and their descriptions.

7.4.2 Analysis of the Features of the Platform

Considering the action performed by each platform functionality, it was possible to put together a list of each functions' purpose in the context of collaboration according to the definitions given in the literature review of this article. During the analysis phase, it was observed that beyond the traditional dimensions of the 3C model, interactivity was present, indicating that this dimension seems to be crucial to understand collaboration in the context of Web 2.0.

The features related to coordination demonstrate lack of communication or interaction between users of the platform. Basically, there is a type of relationship between the user who performed the login and the environment itself. Coordination is directly associated with management activities involving objects arranged on the environment and how the user performs these activities. The coordination features located were: view, create and follow.

On the other hand, cooperation, characterized by the union of manipulation and organization of information, was observed in the action of reporting. This is a group action on the shared space, and the result of the task is to tell the system that there is some kind of activity detrimental to the community. It was noticed that, unlike other features, the action report could only be implemented if there was cooperation between members.

Table 7.2. Description of Local Motors platform's features.

Feature	Description
Preview	This feature is related to the presentation of pages containing ideas, designs, challenges, projects, and files for a project on the platform, i.e. by clicking on the menu option Ideas, for example, the user displays a feed with the ideas that she or he has requested to receive notifications for. From this feed it is possible to access the page of a specific idea. The same happens in the menus of Designs, Challenges and Projects
Create	The Creation functionality refers to the publication of new ideas, designs, projects or challenges on the platform. The user can add a title, a message and/or attachments. This feature can be considered the most important of the platform, since it is from these publications that interactive discussions are generated for the development of new models of vehicle design or collaboration in the improvement of ideas
Follow	From this feature it is possible to monitor the updates on ideas, designs, projects and challenges in order to receive notifications about these items. By following some of the mentioned domains, his/her *feed* is updated
Denounce	This feature allows community members to report ideas, designs, projects or challenges they deem unfit for presentation on the platform
Vote	This feature allows a user to give positive or negative feedback to a particular idea or to vote on a design. From the number of votes it is possible to obtain statistics of the success of a particular item. It is also possible to add votes (positive only) to comments published on pages
Comment	Allows comments on ideas, designs, projects, challenges and topics. In this way, community members can add their opinions and contribute to the development of a car design. For example, based on comments on a previously published design a member of the community can improve this design and publish the changes. It is also possible to attach files to comments
Answer	This feature allows the posting of an answer to a particular comment. Thus, besides the flow of general comments, sub-conversations can be created based on a particular comment
Share	Allows users to share ideas, designs, projects and challenges on social networks (Facebook, Twitter and Google+)
Add discussion topics	Enables the creation of a new forum of discussion on a design, project or specific challenge. Therefore, unlike the comments, which have a general flow of conversation, and occasionally answers to every comment, topics serve to intentionally divide discussions on issues
Contribute	This functionality allows users to download files, computer aided design (CAD) or not, concerning a project and make changes and then add them back to the platform (via comments or forum posts on the project to which that file belongs) so that other members can discuss the changes made
Participate	This feature allows users to submit, within a specified time, their solution to one of the challenges available. If a user does not wish to be a participant, he or she may contribute by sharing ideas and voting on proposed solutions given by other users. There is also a deadline for validation of submissions and subsequently, it is opened the voting period, when the platform members can vote on solutions

Regarding communication, it was observed that in two features there is an exchange of messages and information between users, without necessarily being related to it directly: vote and add discussion topic. The information provided by the act of voting provides to the user a knowledge about the object of his/her review posted, without any kind of relationship with those who did not vote. Meanwhile, the creation of a topic transmits the information to the community in general, without any direct interaction with a member.

Finally, interactivity was seen as a key point for collaboration within the platform. When participants talk about a new idea or project, they are conveying more information than the user who performed the postage of the idea; the focus of the review is a modifiable, manageable and constructive contribution to the network. This aspect is strengthened by the action of answering a comment. The meanings of participation, intervention, bidirectionality and multiplicity are observed in the action to contribute, in which users can collaboratively build the same object. However, as this construction happens asynchronously, since the engineering files (CAD) are typically large, the action cannot be encompassed by the scope of cooperation. Interactivity also occurs in the relationship between the user and the community when participating in the action of a challenge and, finally, in sharing elements of the environment on other networks, increasing their freedom of association.

It should be noted that during the consensus phase some functionalities received a second ranking, based on the result of their actions. They were: the following function (that besides cooperation has a certain degree of interactivity it can be seen as a relationship that results in feedback and comment) and reply to a comment (because the user can transmit information without aggregating any contribution to another user, i.e. without a direct relationship, thus being considered as the communication dimension). However, the first classification is the most suited to the collaborative intent of the functionality within the platform. Table 7.3 summarizes the classification of the extracted features discussed above.

7.5 Final Remarks

Considering the question that guided this research, it can be concluded that, when analyzing the purpose of the features provided by current collaborative platforms, one can see something beyond the 3C model: the interactivity. In the context of the case study analyzed, it does not seem possible to describe the actions performed within collaborative environments in the context of Web 2.0 without this important element.

Even within the current tools, as reported by Fuks et al. [24], which are based only on the 3C model, the fact that interactivity can enhance the volume and quality of production that happens on LMS was not considered. In his work, Fuks and colleagues [24, p. 3–4] described the communication dimension as "providing features

Table 7.3. Classification of Local Motors platform functionality.

Functionality	Classification
1 – View (ideas, designs, projects, challenges, and feed files)	Coordination
2 – Create (ideas, designs, projects and challenges)	Coordination
3 – Follow (ideas, designs, projects and challenges)	Coordination
4 – Report (ideas, designs, projects and challenges))	Cooperation
5 – Vote (ideas and designs)	Comunication
6 – Comment (ideas, designs, projects, challenges and topics)	Interactivity
7 – Reply to comment	Interactivity
8 – Share on social networks (ideas, designs, projects and challenges)	Interactivity
9 – Add Topics Discussion (designs, projects and challenges)	Comunication
10 – Contribute	Interactivity
11 – Attend a challenge	Interactivity

such as forum, chat, instant messages and e-mails"; coordination was seen as the size of the provider features "notifications, evaluation and monitoring of participation." Finally, cooperation has the features bibliography and co-authoring for both teachers and pupils.

It is noted that the co-authoring functionality – key in the context of Web 2.0, where students interact via open social networks site – has no components that strengthen interpersonal relationships, as suggested by Ramaswamy and Gouillart [9]. Co-authoring is seen by Fuks and colleagues [24] just under the operational and functional point of view, regardless of its purpose. It can be extended to capture an underlying social essence, as shown by Silva [22] and Ramaswamy and Gouillart [9].

By assuming the co-authoring functionality under the interactivity dimension, as related to "participation, intervention, bidirectionality and multiplicity of connections" components [27, p. 13], schemes should be leveraged, as it is the case with several features mentioned in the Local Motors platform.

These findings demonstrate the need to carry out further research in this context. Thus, a qualitative study has been proposed as future work. This study would involve managers, who demand these collaborative platforms, and developers who are experts in collaboration and collaborative systems, so that the researcher can confirm the presence of interactivity on current collaborative platforms from the field. This preliminary study indicates that a new model of collaboration is emerging, a model that involves interactivity, communication, cooperation and coordination.

Bibliography

[1] Tapscott, D., Williams, A. D. *Wikinomics*: Como a colaboração em massa pode mudar o seu negócio. Rio de Janeiro: Nova Fronteira S.A., 2006.

[2] O'Reilly, T. *What Is Web 2.0*: Design Patterns and Business Models for the Next Generation of Software, 2005. Available at: http://oreilly.com/web2/archive/what-is-web-20.html. Access: 27 Dec. 2010.

[3] Peng, G., Woodlock, P. The impact of network and recency effects on the adoption of e-collaboration technologies in online communities. *Electron Markets* – Springer 19 (1): 201–210, 2009.

[4] Vreede, G. J., Briggs, R. O., Massey, A. Collaboration engineering: foundations and opportunities. *International Journal of the Association of Information Systems* 10 (3): 121–137, 2009.

[5] Correia-Neto, J. S., Silva, A. A. B., Fonseca, D. *Sites* de Redes Sociais Corporativas: entre o pessoal e o profissional. In: III Encontro de Administração da Informação, 2011, Porto Alegre-RS. *Proceedings* ... Porto Alegre-RS: UFRGS, 2011.

[6] Orlikowski, W. J., Woerner, S. L. Web 2.0: Experimenting with the Connected Web. *CISR MIT Sloan Research Briefing* IX (3), 2009.

[7] Howe, J. (June 2006). The rise of crowdsourcing. *Wired*. Disponível em: http://www.wired.com/wired/archive/14.06/crowds.html. Access: 23 Sep. 2013.

[8] Doan, A., Ramakrishnan, R., Halevy, A. Y. Crowdsourcing Systems on the World Wide Web. *Communications of the ACM* 54 (4): 86–96, 2011.

[9] Ramaswamy, V., Gouillart, F. *The Power of Co-Creation*: build it with them to boost growth, productivity, and profits. New York: Free Press, 2010.

[10] Pimentel, M., Fuks, H., orgs. *Sistemas colaborativos*. Rio de Janeiro: Elsevier, 2011.

[11] Ellis, C. A., Gibbs, S. J. Rein, G. L. Groupware – Some Issues and Experiences. *Communications of the ACM* 34 (1): 38–58, 1991.

[12] Fuks, H., Raposo, A. B., Gerosa, M. A., Pimentel, M. *The 3C Collaboration Model*. In: Khosrow-Pour, ed. The Encyclopedia of E-Collaboration, 2nd edn., 2008. Available at: http://www.groupwareworkbench.org.br. Access: 15 Jan. 2014.

[13] Coleman, D., Antila, D. Enterprise Collaboration: Creating Value Through Content, Context and Process. *Collaborative Strategies*, 2004. Available at: http://www.collaborate.com/. Access: 23 Jun. 2011.

[14] Briggs, R. O., Kolfschoten, G. L., Vreede, G. J., Dean, D. L. Defining Key Concepts for Collaboration Engineering. AMCIS 2006 *Proceedings*. Paper 17, 2006. Available at: http://aisel.aisnet.org/amcis2006/17. Access: 31 Jan. 2013.

[15] Harley, J. J. Collaboration and the Use of Online Collaborative Toolsets in the Project Management Environment. *Thesis*. School of Property, Construction and Project Management. College of Design and Social Context. RMIT University. Australia, 2009.

[16] Fuks, H., Raposo, A. B., Gerosa, M. A., Pimentel, M, Filippo, D., Lucena, C. J. P. Inter- e Intra-relações entre Comunicação, Coordenação e Cooperação. In: SBSC, 27., 2007, Rio de Janeiro-RJ. *Proceedings* ... Rio de Janeiro-RJ: XXVII SBSC, 2007.

[17] OD. Oxford Dictionaries, 2011. Available at: http://oxforddictionaries.com/. Access: 23 Jun. 2011.

[18] Ferreira, A. B. H. *Mini Aurélio*: o dicionário da língua portuguesa. 8th edn. Curitiba: Positivo, 2010.

[19] Lemos, A. *Anjos interativos e retribalização do mundo*: sobre interatividade e interfaces digitais, 2000. Available at: http://www.facom.ufba.br/ciberpesquisa/lemos/interac.html. Access: 02 May 2013.

[20] Kiousis, S. Interactivity: A Concept Explication. *New Media & Society.* (4) 3, pp. 355–383, 2002. Available at: http://www.dtic.upf.edu/~csora/mad/uploads/Main/Spiro_Kiousis_interactivity_2002.pdf. Access: 24 Jun. 2011.

[21] McMillan, S. J. The Researchers and the Concept: Moving Beyond a Blind Examination of Interactivity. *Journal of Interactive Advertising* 5 (2): 1–4, 2005. Available at: http://jiad.org/download?p=58. Access: 24 jun. 2011.

[22] Silva, M. *Sala de aula interativa*, 2nd edn. Rio de Janeiro: Quartet, 2001.

[23] Silva, F. Q. B., Santos, A. L. M., Soares, S. C. B., França, A. C. C., Monteiro, C. V. F., Maciel, F. F. Six years of systematic literature reviews in software engineering: An updated tertiary study. *Information and Software Technology* 53: 899–913, 2011.

[24] Fuks, H., Gerosa, M. A., Raposo, A. B., Lucena, C. J. P. O modelo de colaboração 3C no ambiente AulaNet. *Informática na Educação: teoria & prática.* 7 (1), ISSN 1516-084X, 2004.

[25] Gerosa, M. A., Pimentel, M. G., Filippo, D., Barreto, C. G., Raposo, A. B., Fuks, H., Lucena, C. J. P. Componentes Baseados no Modelo 3C para o Desenvolvimento de Ferramentas Colaborativas. *Proceedings...* Anais do 5th Workshop de Desenvolvimento Baseado em Componentes – WDBC 2005, 07–09 de Novembro, Juiz de Fora-MG, ISBN 85-88279-47-9, 2005, pp. 109–112.

[26] Ferro, G. D. S., Heemann, A. Colaboração em *design* de serviços orientado à otimização dos processos de franquia. *Administração de Empresas em Revista*, 12 (13): 179–191, 2013.

[27] Anderson, C. In the Next Industrial Revolution, Atoms Are the New Bits. *Wired.* Feb 2010.

[28] Magalhães, C. V. C., Santos, R. E. S., Correia-Neto, J. S., Vilar, G. Developing a Social Network of Support to Health Care: The Experience of GenNet. In: *Proceedings of the 18th Brazilian Symposium on Multimedia and the Web*, 2012, pp. 351–354.

[29] Luciano, E. M., Testa, M. G., Rohde, L. R. Gestão de Serviços de Tecnologia da Informação: Identificando a Percepção de Benefícios e Dificuldades para a sua Adoção. *Proceedings...* Anais do XXI EnANPAD, Rio de Janeiro, RJ, 2007.

[30] Yin, Robert K. *Case Study Research* – Design and Methods. USA: Sage Pub, 1989.

[31] Lima, M. O., Pinsky, D., Ikeda, A. A. A utilização do Delphi em pesquisas acadêmicas em administração: um estudo nos anais do EnAnpad. *In*: XI SEMEAD – Seminários em Administração. São Paulo. *Proceedings...* Empreendedorismo emorganizações. São Paulo: FEA-USP, 2008, vol. 11, pp. 1–20.

Ana Lúcia Rodrigues, Carolina Feliciana Machado*, and Ana Paula Ferreira

8 Emotion and Work: an Innovative Relationship?

Abstract: The current chapter aims to uncover the crucial issues and trends regarding workers' emotional intelligence at the workplace. The primary objective of this chapter is to review the studies that correlate emotional intelligence with occupational/job performance. The chapter aims to explore the emotional aspects of intelligence and its related outcomes on employees' performance. Intelligence is considered as an important variable for analyzing workers' capabilities and behaviors necessary to perform a particular task. It is briefly explained what emotional intelligence (EI) is, how it is measured and why it is important in organizations and, specifically, in occupational performance. The case of leadership is also specified.
The chapter reveals interesting findings about the nature of the relationship between individuals' emotional intelligence and their respective performance in the workplace. It also opens the way to the development of innovative management practices. The chapter proposes that emotional intelligence can be used as an approach for attaining organizational results by promoting appropriate worker's behaviors.

8.1 Introduction

In today's rapidly mutable world, employers recognize that people are the key to their success, and that these people require special qualities, primarily those that enable them to survive and sustain themselves in the organization.

The knowledge and skills acquired from past experiences may not be sufficient to come across the new challenges. Intelligent people are those who can retain their knowledge and skills obtained from past experience. That allows them to analyze new situations and develop new solutions. Individuals can solve technical problems far easier than social problems they face in their home, as well as in their professional lives.

Nowadays, organizations have to focus on their employees' emotional intelligence in order to be successful ([1]; Zeidner, Matthews & Roberts, 2009 cited by [2]). That is the reason why the concept of emotional intelligence is explored in order to under-

Ana Lúcia Rodrigues: University of Minho, School of Economics and Management, Department of Management, Campus Gualtar, 4710-057 Braga, PORTUGAL
***Corresponding Author: Carolina Feliciana Machado:** University of Minho, School of Economics and Management, Department of Management, Campus Gualtar, 4710-057 Braga, PORTUGAL, carolina@eeg.uminho.pt
Ana Paula Ferreira: University of Minho, School of Economics and Management, Department of Management, Campus Gualtar, 4710-057 Braga, PORTUGAL

stand different emotions and capabilities people possess and ways to handle them in order to succeed.

Previously, it was believed that there is a positive relationship between people's intelligence quotient (IQ) and their performance, so that intelligent people were perceived to be more successful as compared to less intelligent people. However, IQ ignores some areas like physical aptitude, expertise and other competencies that may result in significant achievements [1]. Emotions should be used to provide a diverse strategy of inspiring and encouraging people. Emotional intelligence brings additional depth to the understanding of human intelligence.

The topic of emotional intelligence has been as controversial as some of the topics in organizational behavior and psychology. However, the expanded and significant role of emotional intelligence in job performance, leadership and other parts of organizational life has increased the validity of this concept [1]. People who are emotionally intelligent are good at recognizing, processing and dealing with their emotions effectively and efficiently. As organizations are spaces where problems emerge and people have to work with each other, emotional information plays a vital role in individual lives (professional, personal and home).

The emotionally intelligent manager can drive under pressure, analyze problems, generate creative solutions, make effective decisions and manage a diverse workforce by helping staff to clarify issues and solve conflicts. The emotionally intelligent salesperson overcomes barriers to achieve goals and to redirect effort for positive results. The emotionally intelligent employee is a team player who feels good about going to work and takes pride in delivering maximum performance on the job.

In the past, emotion was seen as something that interfered with rational and logical thinking. It was immature and messy and to think clearly you needed to stamp out emotion. In the last years, emotion has began to be regarded as adaptive, helpful and functional. Emotions organize our thinking, allow us to know what to pay attention to and motivate our behavior [3].

8.2 Emotional Intelligence (EI)

The topic of emotional intelligence (EI) has garnered several definitions over the last decades. Different researchers have defined Emotional intelligence in different manners. There are many views and definitions of EI, given its scope and complexity.

8.2.1 History and Definition

The idea that emotions and reason are connected has its origins in the writings of Aristotle, who advocates that passions were the motivators of all human behaviors, including the approach and avoidance ones [4]. Actually, it is assumed that EI derives from the concept of social intelligence which was first identified by Thorndike in 1920 [4–7]. According to Roberts and co-workers [4], Thorndike, in defining social intelligence, encompassed the idea of understand and managing the motivations and emotions of men and women, boys and girls, and to act cleverly in human relations. However, the difficulty of empirically distinguishing social intelligence from cognitive intelligence gave rise to the idea that there is no parallel socio-emotional capacity to other measures of intelligence, with two exceptions: the works of Guilford and Gardner.

First, Guilford (1967, cited by [4]) in his intellect model suggested a behavioral category of intelligence, which matches the idea of working with emotional information.

Second, Gardner [8] included intrapersonal intelligence and interpersonal intelligence in his theory of multiple intelligences, both of which are related to emotions. The intrapersonal intelligence refers to the ability to deal with oneself and represent separated arrays of feelings. According to the author, an important dimension of intrapersonal intelligence includes knowledge about the other intelligences. On the other hand, interpersonal intelligence is related to the ability to deal with others, understand them and to know what motivates them. Although social, intrapersonal and interpersonal intelligence aren't named as emotional intelligence, they belong to its realm [6].

Salovey and Mayer (1990, cited by [9]) first used the term 'emotional intelligence' to describe the "capacity to process emotional information accurately and efficiently, including that information relevant to the recognition, construction, and regulation of emotion in oneself and others" (p. 197).

Later, Mayer, Caruso and Salovey [10] proposed that emotional intelligence involves the ability to perceive accurately, appraise, and express emotion; the ability to generate feelings when they facilitate thought; the ability to understand emotion and emotional knowledge; and the ability to regulate emotions to promote emotional and intellectual growth. Grounded on this meaning, they further suggested that emotional intelligence can be divided into four branches, which include: i) perceiving emotions, ii) using emotions to facilitate thought, iii) understanding emotions, and, iv) managing emotions [4, 11]. These levels indicate a growth of complexity of emotional skills: from the first level to the fourth level, from perception to management.

As Roberts and co-workers [4] stated, in 1997, Bar-On characterized EI as "an array of noncognitive (...) capabilities, competencies, and skills that influence one's ability to succeed in coping with environmental demands and pressures" (p. 823). According to Bar-On and co-workers [12] the conceptualization of EI proposed by Bar-On seems to be the most inclusive and comprehensive one. The authors [12] explained that EI includes intrapersonal capacity, interpersonal skills, adaptability, stress management strategies and motivational and general mood factors. In this sense, it embraces the capacity to know and understand one's and others emotions and express feelings and ideas, the ability to solve problems that involve people and to deal with stress and strong emotions and the ability to be optimistic and feel and express positive emotions and feelings [12].

However, it was in the 1990s that the book "Emotional Intelligence" by Daniel Goleman (1995) enhanced the public interest in this subject [4]. Daniel Goleman, the author of several books about EI (including "Working with Emotional Intelligence", 1998), considered that emotional intelligence refers to the ability of knowing one's emotions, managing emotions, motivating oneself, recognizing emotions in others and handling relationships [13]. Goleman (1998, cited by [4]), in his book "Working with Emotional Intelligence" links EI to "competencies associated with self-awareness, self-monitoring, social awareness, and relationship management" (p. 823). Goleman [13] believes that the emotional abilities in relation to cognition abilities are more important in the personal, social and professional success of people. He also believed that emotional intelligences are independent competencies, because they provide unique contributions to job performance, and are necessary but not sufficient to have the capacity to exhibit competencies such as leadership or cooperation [13].

In the scientific literature, all the definitions have received criticisms. The distinction between EI and other constructs such as personality [4] is still not clear. As an example, the study of De Raad [14] that intended to explore to what extent emotional intelligence can be expressed in terms of a standard trait model found that there is a great overlap between the items of the self-reported inventories of EI and the Big Five framework. At the same time, Zeidner and co-workers [7] highlight the lack of agreement in the conceptualization of EI.

Nevertheless, with the dissemination of EI, multiple different research traditions and methods for studying EI emerged in parallel. Different models and measures of EI frequently emerge [4] as will be examined in the next topic.

8.2.2 Measuring Emotional Intelligence

As the theoretical conceptualizations of EI vary, so does the content of the instruments to measure it (Mayer, Salovey et al., 2000a and Mayer, Salovey, & Caruso, 2000b cited by [7]). As Zeidner and colleagues [15] suggested, we can also find little commonality between those instruments.

Among the plethora of models, there are two distinct traditions for measuring EI [7, 16]: self-reported and performance-based measures.

The first tradition measures EI as 'non-cognitive' traits (e.g. assertiveness, optimism, reality testing, conflict resolution), as suggested by Bar-On, measurable by self- and other-ratings. The second tradition positions EI as a cognitive ability measurable by tasks involving cognitive processing of emotional information, and is commonly referred to as 'ability EI'.

It is relevant to examine the seminal work by Bar-On to develop an experimental instrument to measure Emotional Quotient Inventory (EQ-i). According to Bar-On

(2000, cited by [4]), EQ is a pool of emotional and social knowledge that influences our overall activity to cope effectively with environmental demands.

The EQ-i (Bar-On, 1997 cited by [3]; [17]) is a self-reported inventory with 133 items that consists of declarative statements phrased in the first-person singular. Respondents are asked to indicate the degree to which the statement describes themselves, on a five-point scale (*"1=not true of me; 5=true of me"* ([17, p. 798]).

It has five major domains – intrapersonal EI, interpersonal EI, adaptability, stress management and general mood [3] – and these domains are further operationalized into 15 factors [17]. As an example, intrapersonal EI covers emotional self-awareness, assertiveness, self-regard, self-actualisation and independence. Cherniss and coworkers [18], meanwhile, described EQ-i as a "self-report measure with four subscales labeled interpersonal, interpersonal, adaptability, and stress management" (p. 240). Even in the description of the instruments, some inconsistencies are found. Randall [3] also referred to the limitations of the EQ-i, such the coincidence of the instrument with some personality measures and the reliability deficiencies.

In 1997, Cooper and Sawaf proposed another developmental model (limited in its application) [19] and recognized four foundations of EQ using a personal growth approach, which include "emotional literacy, emotional fitness, emotional depth and emotional alchemy" (p. 13).

Following Yunus and Hassan [19], it was Goleman who introduced the performance-based model of EQ. He perceives EQ as embracing a distinct set of abilities that incorporate affective and cognitive skills. The five dimensions of EQ earlier identified by Goleman were further broken down into 25 competencies. According to Goleman [3] they are: i) the self-awareness cluster that includes emotional awareness, accurate self-assessment and self-confidence; ii) the self-regulation cluster that includes self-control, trustworthiness, conscientiousness, adaptability and innovation; iii) the motivation cluster that includes achievement drive, commitment, initiative and optimism; iv) the empathy cluster that includes understanding others, developing others, service orientation, leveraging diversity and political awareness; and, v) the social skills cluster that includes influence, communication, conflict management, leadership, change catalyst, building bonds, collaboration and cooperation, and team capabilities.

The Emotional Competence Inventory (ECI) has its roots in the questionnaire developed by Boyatzis in 1991 and Goleman and Boyatzis, who rewrote items for the non-cognitive competencies [20]. According to the authors [20] the instrument was a useful starting point.

On the other hand, Salovey, Mayer and Caruso (2002 cited by [7]) focused on the ability-based model. They proposed the Multi-factor Emotional Intelligence Scale [7] and its descendant the Mayer–Salovey–Caruso Emotional Intelligence Test (MSCEIT) [3]. In this case, the respondent performs a total of eight tests, two for each of the four branches (the ability to perceive emotion; the ability to use emotion to facilitate thought; the ability to understand emotion; and the ability to manage emotion). Cherniss and colleagues [21] argue that MSCEIT didn't correlate highly with personality or cognitive ability and, in that way, it might be considered that EI is a distinct construct.

Matthews and co-workers [7] suggested that "the ideal EI test should minimally satisfy each of the following four standard psychometric criteria" (p. 182): content validity, reliability, predictive validity and construct validity.

8.3 Job Performance

All organizations are concerned with sustainability and benefits and recognize that organizational benefits depend on individual performance. Therefore, job performance is a central topic for organizations and managers.

Employees performing better will definitely generate outcomes, which primarily include correspondence among employees, quality production and commitment at the workplace [1]. In developed countries, the job description is used to define job responsibilities and performance standards. However, there is an increasing recognition that task performance can't apprehend the full range of job performance [22].

Despite the great relevance of individual performance in organizations and the widespread use of job performance as an outcome measure in empirical research, the concept of job performance still needs to be explained. Nevertheless, during the last 10–15 years, an increasing effort on developing a definition of performance and specifying the performance concept has been made [11].

Not all behavior is subsumed under the performance concept, but only behavior that is relevant for the organizational objectives. Performance is considered as a significant measure, which is associated with the organizational results and success [23]. Performance is related to the impact of an individual's activities over a certain period of time.

Kiyani and colleagues [11]) presented the Campbell, McCloy, Oppler and Sager (1993, p. 40) definition of performance as "what the organization hires one to do and do well". Performance is not described by the action itself but by evaluative procedures, so that only actions, that can be measured are considered to constitute performance (Meyer, Paunonen, Gellatly, Goffin & Jackson, 1989 cited by [11]). Interestingly, individual performance is mainly treated as a dependent variable which makes perfect sense from a practical point of view as individual performance is something organizations want to enhance and optimize [11].

Borman and Motowidlo [22] have indicated that job performance can be divided into task performance and contextual performance. Task performance means that incumbents make a contribution on the technical core of organization through direct productivity operation and material or service support. These activities are related to the specific duty that is required by the organization. Contextual performance is voluntary behavior including helping others or following organizational rules. "Contextual performance is significant to the whole organization, because it supports the organizational, social, and psychological environment in which task performance occurs, promotes communication inside and outside the organization, and relieves tense emotional reaction" [6, p. 1156]. In organizations, people interact with each other and with external constituents (such as customers or suppliers). These interactions are of course characterized by an emotional component that maintains human behavior. Employees with high emotional intelligence are more harmonious and faster integrated into the organization, and achieve higher performance because they can clearly perceive the emotions of colleagues and managers and precisely understand the meaning of other behaviors [6].

As it is known, employees don't only have to accomplish tasks designated by the formal contract between employee and organization, but they are also expected to finish informal tasks. These informal tasks are related to the extra-role behaviors beyond formal role requirements, and are called contextual performance [6]. In order to reach the organizational goals, managing employees' performance is very essential because employees performing better will generate outcomes.

Pulido-Martos and co-workers [2] and Yan-Hong and colleagues [6] stated that employees with high emotional intelligence can better understand the needs of customers and solve problems and conflicts. Compared to employees with low emotional intelligence, they are capable of controlling their own emotions and keeping a positive mood when facing criticism, challenges and stresses from customers.

8.4 Emotional Intelligence and Job Performance

Interest within emotional intelligence has grown dramatically within the past decade. Analysis shows that IQ alone only explains 4–10 percent of accomplishments at work (Sternberg, 1996 cited by [11]). On the other hand, Cherniss and co-workers [21] argue that individuals with high levels of emotional intelligence are more successful in top positions.

Goleman (1995 cited by [4]) postulate that emotional intelligence, which is equivalent, if not more significant than IQ, is a vital measure of success in person's work and personal life.

Day and Carroll [24] found a positive relationship between the perception of emotions and the performance of cognitive decision-making tasks when using the Mayer, Salovey and Caruso Emotional Intelligence Test (MSCEIT). Furthermore, there is some evidence from empirical research to support the idea that emotional intelligence is positively related with task performance. Lam and Kirby [25] applied the Multifactor Emotional Intelligence Scale (MEIS) to test someone's ability of dealing with emotions, and considered that perceiving emotions and regulating emotions contributed more to individual performance than general intelligence.

Sy, Tram and O'Hara [26] tried to examine the relationships among employee's emotional intelligence, manager's emotional intelligence, employee's job satisfaction and performance. The study was conducted with 187 feed service employees from a restaurant franchise, in nine different locations. The hypothesis of the authors was that employee's emotional intelligence was positively associated with job satisfaction and performance and it was corroborated. Thus, the authors concluded that employees with high EI seem to have higher job performance. Some explanations are given, for example employees are more aware of the effect of emotions in their work outcomes and regulate their emotions according to the requirements of the task. However, after controlling for personality factors, the authors didn't meet the traditional standards of significance.

Ali, Garner and Magadley [27] developed a study where the relationship between emotional intelligence and job performance was also explored. In this case, a sample of 310 police officers was taken. The hypotheses of the authors were that there is a positive correlation between EI and job performance and that EI is expected to add incremental validity for predicting job performance beyond cognitive ability and personality traits. It used a self-reported emotional intelligence test and job performance measure tests were implemented. According to the authors, the results show significant correlation between EI levels and police job performance. Justifications for this, given by authors, are that individuals with high levels of EI are more successful in creatively solving problems, completing cognitive tasks and interacting with others than those with lower EI levels. Additionally, it was suggested that after controlling for general mental abilities and personality traits, EI explains an additional incremental variance in predicting police job performance.

Pulido-Martos and colleagues [2] tried to perceive the possible connections between EI and effectiveness during the negotiation process in a sample of 123 workers from different organizations. Their purpose was to examine the incremental validity of perceived emotional intelligence dimensions on effectiveness during negotiation using self-reported measures and controlling for the effect of personality traits. The authors assumed that individuals that deeper understand emotional expressions will be better prepared for negotiation, so will perform better. The results show that emotional regulation accounts for negotiation effectiveness, independently of personality traits. It was suggested that negotiators with higher scores in the emotional regulation component would be more able to create an environment in which their interlocutors do not perceive a power disequilibrium.

In another study developed in a specific context, in this case nursing, Shanta and Connolly [28] consider EI as a crucial component in the nurse's professional role. It is important in terms of the ability of nurses to provide holistic care for patients, peers and themselves. Adopting the four branches of abilities proposed for Mayer and colleagues (already explained in this chapter), Shanta and Connolly [28] concluded the necessity of nurses possessing abilities of EI for their professional practice.

Gondal and Husain [1] who planned a cross-sectional study with 300 employees from the telecom industry analyzed the individual IQ and EI in relation to performance. The results showed that IQ is insignificantly related to an employee's performance and that EI is found to have a significant relationship with employee's performance. Sun (2005, cited by [6]) also suggested that general intelligence and performance had no significant correlation, while emotional intelligence impacted on performance significantly. Not only does employees' emotional intelligence have an effect on task performance, but there is a probable positive relationship between emotional intelligence and contextual performance. Yu and Yuan (2008, cited by [6]) proved that employees' emotional intelligence is positively related with their contextual performance, and leader–member exchange can partially mediate the relationship between managers' emotional intelligence and employees' contextual performance.

Emotional intelligence enhances work performance by allowing people to raise positive relations, perform well in groups and build social assets. Counseling, reinforcement, ability and capability of other people often influence the employee's performance (Seibert et al., 2001, cited by [1]).

EI assists employees in enhancing their performance by allowing them to understand and manage their emotions, cope up efficiently with stress, work well under pressure and prepare for organizational change.

An integrative review and a different perspective are presented by Abraham [29]. Abraham [29] argues that certain emotional competencies (including self-control, resilience, social skills, conscientiousness, reliability, integrity and motivation) are the true predictors of performance and interact with organizational climate and job demands/autonomy to influence performance. The author suggested a model where the relationship between emotional competencies (rather than EI) and performance is intermediated with different organizational variables. Abraham [29] advocates that the replacement of emotional intelligence by emotional competencies as predictors of performance (in a context of positive organizational climate and reasonable job demands) may provide different explanations for the link between those concepts.

By testing five hypotheses, Abraham [29] concluded that "self-control and emotional resilience are considered to delay the onset of a decline in performance from excessive job demands. Social skills, conscientiousness, reliability, and integrity assist to promote trust, which in turn may build cohesiveness among the members of work groups. Motivation may fuel job involvement in environments that promise psychological safety and psychological meaningfulness. A combination of superior social skills and conscientiousness may enhance the self-sacrifice of benevolent employees to heightened levels of dependability and consideration. Finally, emotional honesty, self-confidence, and emotional resilience can promote superior performance, if positive feedback is delivered in an informative manner, and can mitigate the adverse effects of negative feedback" (p. 117).

As suggested, past studies have proven that there was a relationship between EI and work outcomes/behaviors [30].

Employees with higher levels of emotional intelligence are more able to perceive emotions of colleagues and managers and precisely understand the meaning of other behaviors and then carry out adaptive behaviors. As a consequence, they are more pleasant and achieve higher task and contextual performance [6].

Gondal and Husain [1] support the idea that higher EI is related to optimism, stable expressions, and is a facilitator in organizational goal achievement for individuals.

As a conclusion of this topic, we can assume that EI is fundamental to effective performance and provides a base to understand the role of emotions in improving the task and contextual performance. However, it was not always explained how EI was measured or even which definition of EI was taken in the studies presented. Even the way in which job performance was assessed was not clarified: Who evaluated the individual and with what kind of instrument? What kind of performance was evaluated?

8.4.1 The Relationship Between Leadership and Emotional Intelligence

Abraham [29] suggests that emotionally intelligent managers, supervisors and leaders easily deal with their emotions and are more attractive to their peers and associates because they perceive their superiors as emotionally calm colleagues. Kunnanatt [31] argues that the set of characteristics that we may find in emotionally intelligent people may be called EI personality.

Recent research clearly shows that without emotional intelligence a person can have the best training in the world, an incisive, analytical mind, and an endless supply of smart ideas, but still will not become a 'good leader' – some research shows that emotional intelligence is the *sine qua none* of leadership [11].

Today, managers face more challenges than ever and to work successfully with their employees, colleagues and other stakeholders, they need to have great interpersonal abilities [32]. McGee (1996 cited by [19]) highlighted that the failure in leadership is, almost every time, due to poor interpersonal skills. Managers must be able to influence, persuade and negotiate, using great communication skills. To motivate and inspire, a leader needs to be able to reach the hearts and minds of workers and it involves a great understanding of what is important to people and how to involve them in the process [19].

Emotional competencies have become a popular topic among leadership researchers [33]. Ashkanasy and Daus [34] support the notion that the more relational aspects there are in an activity, the more emotional intelligence will be required of the individual who will be put in charge. Thus, leaders who have the ability to perceive their emotions and understand impacts on their actions on those of others, should have a greater probability of providing effective leadership [24].

Rosete and Ciarrochi [35] sought to investigate the relationship between EI, cognitive intelligence and leadership effectiveness. The correlational and regression analyses showed that higher EI was related with higher leadership effectiveness. The authors demonstrated that managers that better comprehend their own feelings and that of their subordinates are more likely to achieve business outcomes, and that they are also considered as well-organized leaders by their employees and direct executives.

Azouzi and Jarboui [36] have published a paper on Corporate Governance Review that explains that the central cause of an organization's problems is CEO emotional intelligence level. Dealing with the relationship between the emotional facet and decision-making processes (specifically, decision biases and effectiveness of the governance mechanisms), the authors concluded that the presence of a high emotional intelligence rate is not always positively correlated with the executive's suggestibility with respect with behavioral biases. Azouzi and Jarboui [36] advanced that it might be important for the well being of individuals, organizations and society as whole, that leaders/CEOs acquire some training in EI.

Extein and co-workers [37] presented the study of Cavallo and Brienza (2004) with more than 300 managers at Johnson & Johnson in which the Emotional Competence Inventory (ECI) was used, a multirater assessment instrument that asks those who work with the individual to rate him/her on a variety of competencies related to EI. The results showed that superior performers scored higher in all four EI clusters (self-awareness, self-management, social awareness and relationship management) based on both superior and subordinate ratings.

Carmeli [38] found a relationship between EI and work attitudes, behaviors and outcomes among senior managers. The results showed that EI enhances positive work attitudes, altruistic behavior and work outcomes, and moderates the effect of work–family conflicts on career commitment but not the effect on job satisfaction.

Chen, Lam and Zhong [39] worked on understanding how the perception of supervisors by their subordinates contributes to high-quality leader–member exchange (that is built gradually, over time, through repeated reciprocal behaviors between the supervisor and the subordinate). Using a longitudinal study on a sample of 285 supervisor–subordinate dyads from a manufacturing firm in China, the authors found that supervisor-rated emotional intelligence of subordinates predicts the quality of leader–member exchange and that leader–member exchange positively predicts work performance. Chen and his colleagues [39] concluded also that leader–member exchange mediates the interactive effect of EI and trust in the supervisor on work performance.

The study of Yunus and Hassan [19] found significant relationships between EI and employee's current job performance. Yunus and Hassan [19] contend that organizations should use an emotional intelligence test as a developmental instrument to identify and promote potential leaders. According to them, managers should be promoted when they have strong emotional intelligence competencies, in parallel with technical and educational skills. Those ingredients are quite important for their success.

Two other studies from Bar-On, Handley and Fund (2005, cited by [21]) looked at EI and performance in military environments. One of the studies was conducted in the US Air Force to see if EI assessment could help predict performance in military recruiters.

The study measured EI using the EQ-i, and performance ratings were based on individual productivity. Another study looked at EI, as measured by the EQ-i, and performance, as measured by peer nomination, criterion group membership, and commander evaluations in the Israeli Defense Forces. Both studies found military recruiters and combat soldiers considered high performers had significantly higher scores on the EI measures than low performers.

A study of Cavazotte, Moreno and Hickmann [40]) investigated the effects of intelligence, personality traits and emotional intelligence on transformational leadership and the performance of leaders in the organizational context. Using 124 midlevel managers from a Brazilian company from the energy sector as a sample, data were collected by evaluating the personality, the emotional intelligence, the intelligence, the leadership traits of the managers and the transformational leadership (using the subordinates of the managers). The findings of the authors suggest that leadership effectiveness (measured as the achievement of the organizational results) is a direct function of the transformational behaviors of leaders and an indirect function of individual differences (such as intelligence) that work through transformational behaviors. When EI is isolated and related with transformational leadership, the effect became nonsignificant. So it is not possible, according to those findings, to assume a direct relationship between EI and leadership effectiveness.

Kiyani and his colleagues [11] stated that there is support for the idea that managers' emotional intelligence positively accounts for differences in employee outcomes. Studies show that emotional intelligence is positively related to employee's performance (Higgs, 2004 cited by [11]). Wong and Law (2002 cited by [11]) also found that the emotional intelligence of managers has a causal effect on the job performance and organizational citizenship behavior of their subordinates. The failure of leaders in the workplace is due to underdeveloped emotional intelligence. Having emotional intelligence will equip a leader with skill to manage people. Many business leaders agreed that success in the workplace is strongly influenced by personal qualities such as perseverance, self-control, and skill in getting along with others.

Bunker and Wakefield [32] claim that managers face more difficulty than ever in working effectively with their employees, colleagues, and other stakeholders. Yunus and Hassan [19] support the idea that the failure of leaders in the workplace is due to underdeveloped soft skills and lack of emotional intelligence. Having emotional intelligence and soft skills will equip a leader with skills to manage people, specifically inter and intrapersonal relationships successfully. Therefore, effective leadership requires emotional skills so that those who aspire to become a leader are not likely to succeed without highly developed skills in these areas.

As Bar-on concludes, "one's ability to succeed in coping with environmental demands and pressures is a function of one's emotional intelligence. One of the reasons people leave organizations is because of a poor relationship with their boss. EI requires that we learn to acknowledge and value feelings in ourselves and others – that appropriately respond to them, effectively applying the information and energy of emotions in our daily life and work" [19].

8.5 EI Relevance for Management

Following Waterhouse [41], there is much evidence of the link between EI and a variety of outcomes, particularly in the workplace. As suggested by Yunus and Hassan [19], emotional intelligence can impact everyone in the organization, raising the level of performance.

The breadth of connections that emotional intelligence develops with other organizational variables deserves to be considered by management. Some theorists claim that EI is an important predictor of all areas of workplace performance.

The orientation for results, the positive effects on the processes of decision making and positive attitudes towards work, the enhancement in performance and the capacity for being adaptable should be considered as important contributes for organizational success.

The study of Day and Carroll [24] tried to demonstrate the importance of EI in cognitive decision making, arguing that emotional perception predicted individual performance on the proposed task. The authors also suggested that individuals with high levels of EI are able to take advantage of their emotions by using them to facilitate reasoning, creative thinking and decision making.

Job satisfaction must also be considered. Sy and colleagues [26] suggest that employees with higher EI have higher job satisfaction, and suggested that those employees are more adept at identifying and regulating their emotions. The authors argue that employees with high levels of EI are more able to search for the causes of their stress, thereby enabling them to develop new coping strategies and ways of managing their emotional reactions to the stressors. Similar results were found for job performance.

Mostly intelligent people who have a bright academic record sometimes are not good in social interactions and interpersonal dealings. It does not indicate that IQ should be ignored yet it indicates that EI is a more important construct than IQ for enhancing organizational effectiveness. As it was suggested earlier in this chapter, the high performance of workers and the success of their learning seem to result from a synergic effect of emotional and rational capabilities.

At the same time, leaders tend to be more effective, in terms of pursuing objectives/results and satisfaction, as they have higher degrees of emotional intelligence. The findings of Rosete and Ciarrochi [35] cited early in this chapter, suggest that executives with higher levels of EI are more expected to achieve business outcomes and be considered effective. On performance management, the authors argue that, in order to deliver feedback to the employees, EI is useful for leaders in identifying who performs complex tasks well and who deals effectively with colleagues and staff.

Leaders with great emotional competencies promote a creative environment for workers, are more successful in conflict management and get more from their teams. Emotional intelligence may help the managers to develop employees in terms of positive and committed workforce by developing and enhancing their emotional capabilities. Emotional intelligence may indeed be a key determinant of employees' effective performance at the workplace.

In increasingly complex environment contexts, organizations must be aware of leaders and employees that can withstand the economic fluctuations and even to take advantage of any opportunities arising from these research and development efforts, competitor advances and globalizations. This means that management have to constantly re-engineer a systematic employment process or an approach that can bring forth the desired organizational performance and the workforce that can deal with changes and not rally against them. The benefits of emotionally intelligent organizational members might influence positive outcomes with regard to organizational performance and effectiveness.

Any organizational actor calls for their emotional management throughout their professional work. As we tried to demonstrate, emotions are part of the organizational setting and emotional intelligence should be considered in modern times. At a time when organizations are striving for competitiveness and its maintenance in a 'league of honor', where competitive advantage is the key requirement, to ensure the quality of service requires a great effort by those dealing directly with clients and, on the other hand, requires leadership that goes far beyond the exercise of authority.

8.6 Conclusions

The latest researches in organizations have indicated new evaluation criteria of the human being, it does not matter just how smart the individuals are, intellectually speaking, nor the training or expertise, but mainly the way they deal with themselves and with others. They also indicate that human abilities are inherent to professional success, because of the excellence in work quality, particularly for leadership positions.

Human resources management have to constantly re-engineer a systematic employment process or an approach that can bring forth the desired workforce and leaders that can work with and take advantage of changes. In management and research framework, the groundbreaking nature of the relationship between EI and work should be emphasized. New employees with high emotional awareness and regulation should be hired and retained. As was exposed in this chapter, emotional intelligence may indeed be a key determinant of employees' effective performance at the workplace. Emotional intelligence may help managers to develop employees in terms of a positive and committed workforce by upgrading and enhancing their emotional capabilities.

Assuming EI as a central concept in human resource management practices, it is important to understand that the organizational outcomes will be achieved through an emotionally intelligent and effective leadership and a workforce that displays appropriate behaviors, from the emotional and rational point of view. In a pioneering way, the relationship between two important organizational variables, performance and emotional intelligence, was explored. However, as a consequence of those findings, some new management practices should be studied and defined.

In order to reinforce and explain this innovative relationship between EI and work in the future it will be important to clarify and distinguish the conceptualization of emotional intelligence and the way it can be measured. Although the results of the studies presented reveal some consistent relations between EI and performance (task and contextual performance of employees or leaders), it is still not well established that when we are talking about EI, it is a different construct of personality. It means that, the overlap between EI and other constructs should be overcome.

It would also be important to develop some comparative studies, by comparing different cultures, and even different measures with the same people. To conclude, it would be crucial to know and understand how job performance or leadership is evaluated. Since performance evaluation is considered a controversial practice in organizations, it would be interesting in future studies to explicitly clarify how the assessment is conducted and how the notion of performance is conceptualized.

Bibliography

[1] Gondal, U., Husain, T. A Comparative Study of Intelligence Quotient and Emotional Intelligence: Effect on Employees' Performance. *Asian Journal of Business Management* 5(1), 153–162, 2013.

[2] Pulido-Martos, M., Lopez-Zafra, E., Augusto-Landa, J. Perceived emotional intelligence and its relationship with perceptions of effectiveness in negotiation. *Journal of Applied Social Psychology* 43, 408–417, 2013.

[3] Randall, K. Emotional intelligence: what is it, and do Anglican clergy have it? *Mental Health, Religion & Culture* 1–9, 2013.

[4] Roberts, R., MacCann, C., Matthews, G., Zeidner, M. Emotional Intelligence: Toward a Consensus of Models and Measures. *Social and Personality Psychology Compass* 4(10), 821–840, 2010.

[5] Whiteoak, J., Manning, R. Emotional intelligence and its implications on individual and group performance: a study investigating employee perceptions in the United Arab Emirates. *The International Journal of Human Resource Management* 23(8), 1660–1687, 2011.

[6] Yan-Hong, Y., Run-tian, W., Wang, K. The Influence of Emotional Intelligence on Job Performance: Moderating Effects of Leadership. International Conference on Management Science & Engineering (16th), 14–16 September, 1155–1160, 2009.

[7] Zeidner, M., Matthews, G., Roberts, R. Emotional Intelligence in the Workplace: A Critical Review. *Applied Psychology: An International Review* 53(3), 371–399, 2004.

[8] Gardner, H. The Theory of Multiple Intelligences. *Annals of Dyslexia* 37, 19–35, 1987.

[9] Mayer, J., Salovey, P. Emotional intelligence and the construction and regulation of feelings. *Applied and Preventive Psychology* 4, 197–208, 1995.

[10] Mayer, J., Caruso, D., Salovey, P. Emotional intelligence meets traditional standards for an intelligence. *Intelligence* 27(4), 267–298, 2000.

[11] Kiyani, K., Saher, N., Saleem, S., Iqbal, M. Emotional Intelligence (EI) and Employee Outcomes: The Mediating Effect of Authentic Leadership Style. *Interdisciplinary Journal of Contemporary Research in Business* 5(1), 394–405, 2013.

[12] Bar-ON, R., Brown, J., Kirkcaldy, B., Thomé, E. Emotional Expression and Implications for Occupational Stress; an implication of the Emotional Quotient Inventory (EQ-i). *Personality and Individual Differences* 28, 1107–1118, 2000.

[13] Goleman, D. *Trabalhar com Inteligência Emocional.* Lisboa: Temas e Debates – Círculo de Leitores, 2012.

[14] De Raad, B. The Trait-Coverage of Emotional Intelligence. *Personality and Individual Differences* 38(3), 673–687, 2005.

[15] Zeidner, M, Matthews, G. M., Roberts, R. Slow Down, you Move too fast: Emotional Intelligence Remains an "elusive" intelligence. *Emotions* 1, 265–275, 2001.

[16] Mayer, J., Roberts, R., Barsade, S. Human abilities: Emotional intelligence. *Annual Review of Psychology* 59, 507–536, 2008.

[17] Dawda, D., Hart, S. Assessing emotional intelligence: reliability and validity of the Bar-On Emotional Quotient Inventory (EQ-i) in university students. *Personality and Individual Differences* 28, 797–812, 2000.

[18] Cherniss, C. et al. Process-Designed Training: a new approach for helping leaders develop emotional and social competence. *Journal of Management Development* 29(5), 413–431, 2010.

[19] Yunus, N., Hassan, C. The Importance of Emotional Intelligence and Soft Skills in the Workplace. *International Conference on Innovation, Management and Technology Research*, Malacca, Malaysia, 21–22 May 2012.

[20] Boyatzis, R., Goleman, D., Rhee, K. Clustering Competence in Emotional Intelligence: insights from the emotional competence inventory. In Bar-On, R., Parker, J. D. A., eds. *Handbook of emotional intelligence*. San Francisco: Jossey-Bass, 343–362, 2000.

[21] Cherniss, C., Extein, M., Goleman, D., Weissberg, R. Emotional Intelligence: What does the research really indicate? *Educational Psychologist* 41(4), 239–245, 2006.

[22] Borman, W., Motowidlo, S. Task Performance and Contextual Performance: The Meaning for Personnel Selection Research. *Human Performance* 10(2), 99–109, 1997.

[23] Wall, T., Michie, J., Patterson, M., Wood, S., Sheehan, M., Clegg, C., West, M. The Validity of Subjective Measures of Company Performance. *Personnel Psychology* 57, 95–118, 2004.

[24] Day, A., Carroll, S. Using an ability-based measure of emotional intelligence to predict individual performance, group performance, and group citizenship behaviours. *Personality and Individual Differences* 36, 1443–1458, 2004.

[25] Lam, L., Kirby, S. Is Emotional Intelligence an Advantage? An Exploration of the Impact of Emotional and General Intelligence on Individual Performance. *The Jounal of Social Psychology* 142(1), 133–143, 2002.

[26] Sy, T., Tram, S., O'Hara, L. Relation of employee and manager emotional intelligence to job satisfaction and performance. *Journal of Vocational Behavior* 68, 461–473, 2006.

[27] Ali, O., Garner, I., Magadley, W. An exploration of the relationship between emotional intelligence and job performance in police organizations. *Journal of Police and Criminal Psychology* 27, 1–8, 2012.

[28] Shanta, L., Connolly, M. Using King's Interacting Systems Theory to Link Emotional Intelligence and Nursing Practice. *Journal of Professional Nursing* 29(3), 174–180, 2013.

[29] Abraham, R. Emotional Competence as Antecedent of Performance: A contingency framework. *Genetic, Social, and General Psychology Monographs* 130(2), 117–143, 2004.

[30] Prati, L., Douglas, C., Ferris, G., Ammeter, A., Buckley, M. Emotional Intelligence, Leadership Effectiveness, and Team Outcomes. *The International Journal of Organizational Analysis* 11(1), 21–40, 2003.

[31] Kunnanatt, J. Emotional Intelligence: theory and description. A competency model for interpersonal effectiveness. *Career Development International* 13(7), 614–629, 2008.

[32] Bunker, K., Wakefield, M. Now More Than Ever, Soft Skills Are Needed. *Leadership in Action* 24(1), 17–21, 2004.

[33] Gooty, J., Connelly, S., Griffith, J., Gupta, A. Leadership, affect and emotions: A state of the science review. The Leadership Quarterly, 17(1), 5–17, 2010.

[34] Ashkanasy, N., Daus, C. Rumors of the death of emotional intelligence in organizational behavior are vastly exaggerated. *Journal of Organizational Behavior* 26(4), 441–452, 2005.

[35] Rosete, D., Ciarrochi, J. Emotional intelligence and its relationship to workplace performance outcomes of leadership effectiveness. *Leadership and Organizational Development Journal* 26(5), 388–399, 2005.

[36] Azouzi, M., Jarboui, A. CEO emotional intelligence and board of director's efficiency. *Corporate Governance* 13(4), 365–383, 2013.

[37] Extein, M., Cherniss, C., Goleman, D., Weissberg, R. P. Emotional Intelligence: What does the research really indicate?. *Educational Psychologist* 41(4), 239–245, 2006.

[38] Carmeli, A. The relationship between emotional intelligence and work attitudes, behavior and outcomes: an examination among senior managers. *Journal of Managerial Psychology* 18(8), 788–813, 2003.

[39] Chen, Z., Lam, W., Zhong, J. Effects of perceptions on LMX and work performance: effects of supervisors' perception of subordinates' emotional intelligence and subordinates' perception of trust in the supervisor on LMX and, consequently, performance. *Asia Pacific Journal of Management* 29, 597–616, 2012.

[40] Cavazotte, F., Moreno, V., Hickmann, M. Effects of leader intelligence, personality and emotional intelligence on transformational leadership and managerial performance. *The Leadership Quarterly* 23, 443–455, 2012.

[41] Waterhouse, L. Multiple Intelligences, the Mozart Effect, and Emotional Intelligence: A Critical Review. *Educational Psychologist* 41(4), 207–225, 2006.

Index

Bei Fragen zur Produktsicherheit wenden Sie sich bitte an:
If you have any questions regarding product safety,
please contact:

Walter de Gruyter GmbH
Genthiner Straße 13
10785 Berlin
productsafety@degruyterbrill.com